INTRODUCTION TO BOOLEAN ALGEBRA AND LOGIC DESIGN

A PROGRAM FOR SELF-INSTRUCTION

INTRODUCTION TO BOOLEAN ALGEBRA AND LOGIC DESIGN

GERHARD E. HOERNES

MELVIN F. HEILWEIL

McGRAW-HILL BOOK COMPANY

New York San Francisco Toronto London

INTRODUCTION TO BOOLEAN ALGEBRA AND LOGIC DESIGN

Library of Congress Catalog Number 64-16484

ISBN 07-029183-7
ISBN 07-029185-3

12 13 14 15 VBVB 7 5 4 3 2

TO GERIANNE
AND
TO VICTOR

PREFACE

The purpose of this book is to introduce the algebra used in the logical design of computers. Primarily intended for the education of the logic designer, it is sufficiently general to be used by those having only an academic interest in the subject. The book is based on the application of logic design to electronic switching circuits, but Appendix D contains sufficient information for applying the material to relay switching circuits. An understanding of electronics is helpful but not a prerequisite. Appendix C contains a short description of typical circuits. A general understanding of elementary algebra is a prerequisite.

The subject matter of this book has been divided into two major parts. Part I introduces Boolean algebra, proves theorems, shows how algebraic manipulations can be performed, introduces logic blocks, and gives a method for translating word problems into Boolean algebra. Part II introduces formal minimization techniques (the Quine-McCluskey and the Karnaugh map methods) for determining the optimum design of a logic circuit.

This book has been written as a text for a formal course in logic design, but its self-instructional format makes it equally useful for individual study. Problems have been included after each chapter, with answers given in the back of the book. Many of the problems in Part II illustrate practical applications of the material to the computer field.

The authors are grateful to the many individuals who made possible the publication of this book. In particular, the authors wish to thank Mr. Gerald A. Maley for technical reviewing; Messrs. Richard C. Marra and Harvey S. Long for assisting in the writing, production, and distribution of the test version; Mr. Thomas F. Cummings for initially suggesting the writing of self-teaching material by the authors; and IBM management for permitting the use of the material in the various educational programs.

G. E. Hoernes

M. F. Heilweil

HOW TO USE THE BOOK

Except for introductory and background material, this book has been written in self-teaching form. This means that the information has been broken down into easily understandable units of information, called *frames,* each containing a single fact or concept to be learned. In addition to the information, the frame contains a question whose answer is given below the frame. The purpose of the question and answer is to determine whether you understand the information before continuing on to the next frame. Most answers include an explanation to aid you if your answer is wrong, or if you are not sure of it. If you still do not understand, you should review the frame or previous frames until you do. If your answer is correct, you may skip over the explanation.

After reading a question, *it is important to write down your answer* before comparing it to the correct answer. Not only will writing down an answer help you to remember it, but it will also keep you mentally honest by forcing you to crystallize a vague idea into a definite answer. Your answer may be written in the book or on a separate sheet, but it is easier to write it in the book when there are forms or tables to bc completed.

Immediately above each question and each answer, there is a heavy line and a frame identification number. The question part of the frame will have a number such as 23Q, where Q stands for Question, and the answer part of the same frame will be identified by 23A, where A stands for Answer. When reading a question, it may be helpful to cover the answer with a card or working pad until you have completed your own answer.

Whenever possible, it is best to read a complete chapter in one sitting. If you must stop before the end of a chapter, a good stopping point is at a frame which summarizes the previous material because you can then reread this summary when starting again. At the end of each chapter you will find practice problems which will serve as an additional test of how well you understand the chapter and whether you need additional review. The answers to the problems are at the back of the book.

An important characteristic of the self-teaching method is that you are free to set your own pace in learning the material, and you may slow down or speed up that pace whenever necessary.

Summary

1. Read the statement and question while keeping the answer covered.

2. Write down your answer and then compare it with the correct answer.

3. You may skip the explanation if your answer is correct.

4. If your answer is incorrect and the explanation does not help you, reread the frame or previous frames.

CONTENTS

PART I

BOOLEAN ALGEBRA

The first chapter of this part describes the purpose of Boolean algebra in computers and demonstrates its value. The next four chapters introduce the theorems of Boolean algebra and the truth table and take the student through some algebraic manipulations, preparing the reader for minimization techniques which are taught in Part II. The last chapter in Part I shows how word statements can be translated into Boolean algebra.

Chapter 1

INTRODUCTION

The purpose of this chapter (which is written in ordinary text form) is to introduce the subject of logic design and to show the need for Boolean algebra. This is done by describing how information may be represented using two-level, or binary, signals, and how combinations of AND, OR, and INVERT logic blocks may be used with these binary signals to perform the various functions in a computer. A simple example is chosen to illustrate how a function may be specified and how a logic circuit is designed to perform that function. Finally, it is shown that other logic circuits will perform the same function and that a method is needed for finding the best circuit to do the job.

Information Representation

The logic designer builds a computer by starting with a few types of simple, basic circuits, called *logic blocks*, which he then interconnects in large quantities to perform the various functions of the computer.

Before getting into logic design, it is important to understand how information (numbers, control signals, etc.) is represented in a digital computer as compared to an analog computer.

In an analog computer, information is represented by signals whose amplitude may have any value (within a fixed range). An analog signal may be used to represent a numerical value by making the amplitude of the signal analogous to a number. For example, a voltage of 7.28 volts may represent the number 7.28 (or 0.728, or 72.8, etc.), and a voltage of 3.61 volts may represent the number 3.61 (or 0.361, or 36.1, etc.). The accuracy of this type of number representation depends upon the accuracy with which the voltages can be generated and measured, and it is therefore difficult to accurately represent numbers of more than three or four decimal digits. For this reason, analog computers are not used where great accuracy is required.

To represent a number in digital form, the voltage is not allowed to take any value but is limited to a set of values separated from each other by forbidden bands. For example, the voltage on a wire may be limited to 10 discrete levels, which are sufficient to represent one decimal digit. To represent numbers having more than one digit, more than one wire is used. Assuming that the circuits are operating properly, there is no doubt about the value of any digit, since the indeterminancy of a measuring device is not a limiting factor. The almost unlimited accuracy of digital computers, which results from the use of digital signals, is the main reason for their extensive use. A price must be paid for this accuracy, however, and digital computers are generally found to be larger and more expensive than analog computers.

The previous example of a digital system was not very realistic in assuming that ten levels are permitted on a single line. In almost every digital computer, the number of allowable levels on a line is limited to two. One reason for this choice is that a vacuum tube or transistor is much easier to operate when completely *on* or completely *off* (the conditions that define the two levels) than to operate it in a larger number of intermediate levels. Another reason is that two levels can be distinguished more easily than a larger number of levels within the same voltage range. The result is that two-level circuits are simpler, less expensive, and more reliable than multilevel circuits, although more two-level than multilevel circuits are needed for any application.

Signals which can have only two levels are called *binary signals*. The actual values of these levels depend to a great extent on the circuit components. For example, vacuum-tube circuits often use the levels of -10 and +30 volts, while transistor circuits often use such levels as 0 and +12 volts, -3 and +3 volts, etc. It should be mentioned that, in practice, a voltage level is often specified within rather wide limits and is more a voltage band than a precise voltage. For example, a +3 volt level may actually be a voltage band of from +2 to +4 volts. When a particular level is referred to, it should be thought of as a nominal value, i.e., representative of a band of voltages which are treated as a single level.

A logic designer does not need to concern himself with the actual voltages, and he therefore makes use of general terms to describe the levels. For example, he might call them the *up* level and the *down* level, or the (relatively speaking) *positive* level and *negative* level. Various symbols have also been used to describe these levels, a common choice being the symbols + and -. These symbols do *not* refer to positive and negative voltages with respect to ground, but rather they refer to the *more* positive and *more* negative voltages with respect to each other. If the two levels are -12 and -6 volts, the -12-volt level would be called ***minus*** (-) and the -6-volt level would be called ***plus*** (+). Another common set of symbols, and the ones used in this book, are the symbols 0 and 1. Unfortunately, these same symbols are used to represent numbers, and usually the only way to distinguish between the two meanings is by the context in which they are used. In this book, the symbol 1 will be used for the more positive voltage level, while 0 will be used for the more negative voltage level.

Representing Information with Binary Signals

All information within a digital computer is placed in the form of these binary signals. Let us first consider how numbers can be represented in this form. Each decimal digit has 10 (0-9) possible values, but since the voltage on a wire can have only two values, a single wire is not sufficient to represent a decimal digit. Using 10 wires, each wire can stand for one of the 10 values of the digit. If one and only one of 10 wires is allowed to be at the more positive or 1 level at any time, while the other nine are at the more negative or 0 level, then the value of the digit can be identified by observing which wire is at the 1 level. For example, when wire 8 is at the 1 level, the value of the decimal digit is 8. The following table shows the signal levels which represent some of the digit values.

Digit values	Wire number									
	0	1	2	3	4	5	6	7	8	9
0	1	0	0	0	0	0	0	0	0	0
1	0	1	0	0	0	0	0	0	0	0
....	...	...	...	...	...	...	...	...	...	...
8	0	0	0	0	0	0	0	0	1	0
9	0	0	0	0	0	0	0	0	0	1

Notice that this method takes 10 wires to represent just *one* digit, 20 wires to represent a two-digit number and so on.

There is another way to represent a decimal digit that requires only four wires. If the signals are *not* restricted to only one wire being at the 1 level, then there are 16 different combinations of signal levels that may be present on the four wires. If 10 of these combinations are assigned decimal values, and if the remaining six combinations are not allowed to occur, then the four wires can be used to represent a decimal digit.

Digit values	Wire number			
	8	4	2	1
0	0	0	0	0
1	0	0	0	1
2	0	0	1	0
3	0	0	1	1
4	0	1	0	0
5	0	1	0	1
6	0	1	1	0
7	0	1	1	1
8	1	0	0	0
9	1	0	0	1
	1	0	1	0
Not	1	0	1	1
allowed	1	1	0	0
	1	1	0	1
	1	1	1	0
	1	1	1	1

To represent a five-digit number such as 76,543, fifty wires are needed for the first method, while only 20 wires are needed for the second method. It is, of course, possible to use fewer wires if digits are to be transmitted one at a time instead of all at once.

So far we have discussed only decimal numbers. There is another number system, called the ***binary number system***, in which each digit can have only two values (0 and 1) instead of 10 values. This system, which is discussed later in the book (see Appendix B—The Binary Number System), has the advantage that each digit of a number can be represented on a single wire. Many computers use binary numbers because of this advantage, and also because arithmetic operations are simpler in binary than in decimal.

Letters of the alphabet can also be represented in a computer by using binary signals. Since there are 26 letters, five wires would be needed, and six of the 32 possible combinations of signal levels on the five wires would not be allowed. If six wires are used, there would be enough combinations (64) to represent all letters, numbers, and commonly used typographical symbols.

In addition to numbers and letters, control information is also represented with binary signals. In general, there are two types of control information. The first type specifies the operation (addition, subtraction, multiplication, etc.) to be performed by the computer, and these can be represented on a set of wires by assigning to each operation a unique combination of signals. For example, if there are eight possible operations, then each of the eight combinations of signals on three wires would be assigned to an operation. Addition might be identified by 100 while subtraction might be identified by 101.

The second type of control information is used to control the performance of each operation. This mainly consists of determining when information is to be sent from one location to another. For example, a control signal on a wire may determine when a number is to be sent from the place where the number is stored to a place where it is to be added to another number. A control signal on another wire may determine when a different number is to be sent from the place where it is stored to a place where it can be displayed on a row of lights. The signals which are used to control these transfers are binary signals.

In summary, *all* types of information in a digital computer are represented using binary signals. This is important to remember because all the logic techniques in this book are based on the use of binary signals. In fact, the study of logic design is essentially aimed at the design of any equipment which represents information in this manner.

Logic Blocks

Logic blocks, which are the basic circuits of a computer, combine and manipulate the binary signals so as to perform most of the functions of the computer. Let us now take a closer look at these logic blocks.

A logic block may have one or more input lines, but only one output line. A logic block having three inputs is drawn as follows:

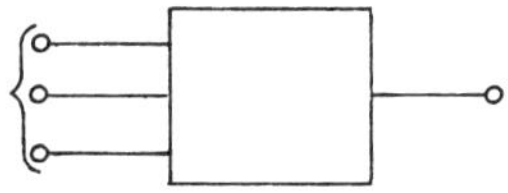

Since there are different kinds of logic blocks, the name or symbol of the block will be placed in the box as identification.

A logic block is defined by giving the output level for every possible combination of input levels.

Note that a logic block is defined in terms of its input-output characteristics and *not* by a circuit description. This is because a logic block should not be considered as a specific circuit, but rather as a "black box" having a well-defined set of input-output characteristics. It is therefore possible to have a number of different circuits which can all be considered as the same kind of logic block.

Three types of logic blocks will be introduced in this chapter. They are the AND block, the OR block, and the INVERTER block. These three blocks are perhaps the most common found in present-day computers and should serve as a very practical basis for the study of logic design.

1. AND block

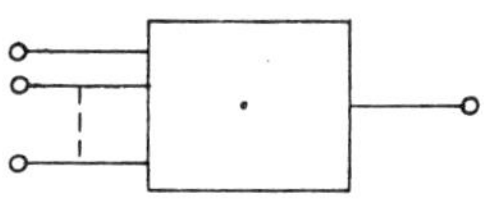

The AND block, which is identified by the symbol ·, may have any number of inputs. Its output is at the 1 level only if *all* inputs are at the 1 level. If any input is at the 0 level, then the output will be at the 0 level. This logic block is called an AND block because its output is 1 only if the first input *and* the second input *and* the third input, etc., are 1.

2. OR block

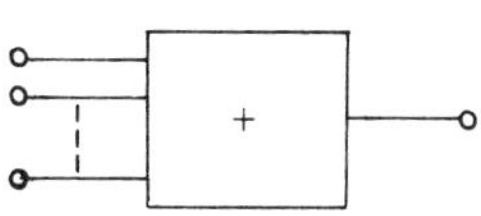

The OR block, which is identified by the symbol +, may have any number of inputs. Its output is at the 1 level if *any* input is at the

1 level. If all inputs are at the 0 level, then the output will be at the 0 level. This logic block is called the OR block because its output is 1 if the first input *or* the second input *or* the third input, etc., is 1.

3. INVERTER block

The INVERTER block, which is identified by the symbol I, has only a single input. Its output is at the 1 level if its input is at the 0 level. Its output is at the 0 level if its input is at the 1 level. This logic block is called the INVERTER block because its output can be considered the *inverse* of its input.

The major part of this book is devoted to methods for determining how these blocks should be interconnected to perform any function required in a computer.

Circuits

Ideally, the logic designer need not concern himself with the particular circuits used for the logic blocks, but in reality this is never possible. One important reason is that a logic designer often has to choose one of a few possible designs, and his choice usually depends upon their cost. Since the cost of each design depends upon the particular circuits used, it is generally impossible to determine the least expensive design unless the circuits are known. In order to make this book more practical, three circuits will be introduced for later use in examples and problems. The two circuits used for the AND and OR blocks are called *diode* logic circuits because the main component in each is the diode (either a vacuum tube or semiconductor diode). The circuit used for the INVERTER block will be shown in a triode vacuum-tube version and also a transistor version. For the purposes of this book, it is *not* necessary to understand the internal operation of these circuits. It is only necessary to know how many diodes and transistors (or vacuum tubes) each circuit requires. The diode and transistor counts will be used as an approximation of the cost of each design. When necessary, the cost of a transistor may be assumed to be three times the cost of a diode. This 3 to 1 ratio is only an arbitrary figure, since the actual ratio of costs depends on many factors which are continually changing.

For an explanation of the operation of the following circuits, and for an explanation of the circuit symbols, see Appendix C.

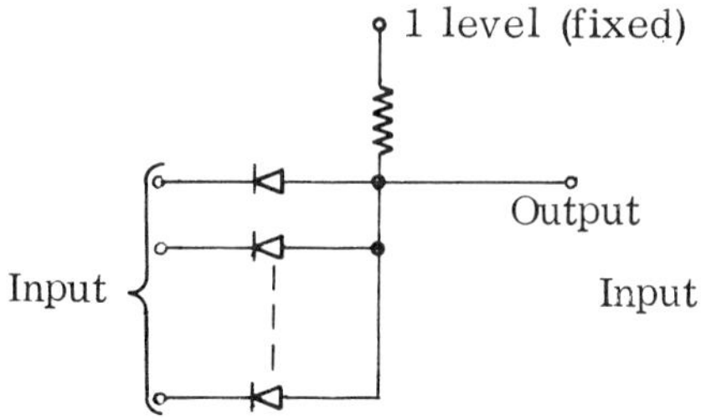

Diode AND

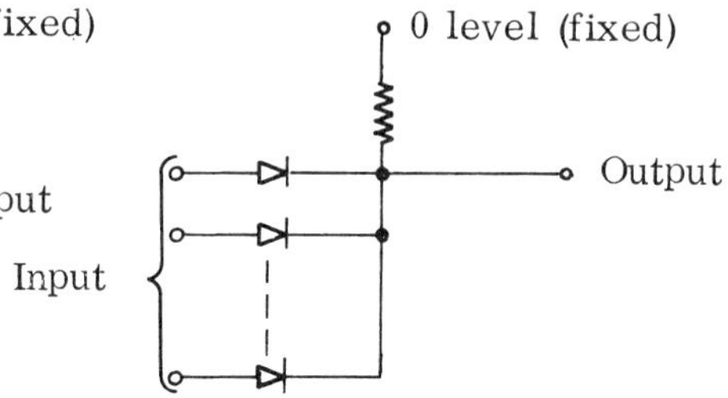

Diode OR

The two circuits differ in the way the diodes are connected (the "arrows" are in different directions) and by the fixed voltage levels connected to the resistor. It can be seen from these diagrams that the "cost" of either circuit is one diode per input.

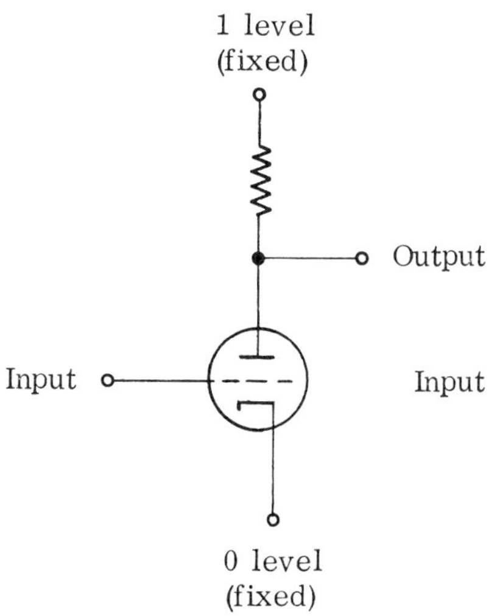

Vacuum-tube
INVERTER

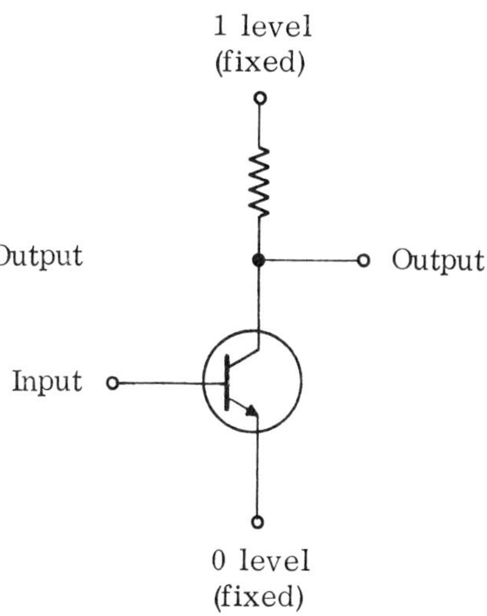

Transistor
INVERTER

The "cost" of an INVERTER is one triode vacuum tube or one transistor.

Having decided on a specific set of circuits for our logic blocks, we can now summarize the cost of each logic block.

Block — **Cost**

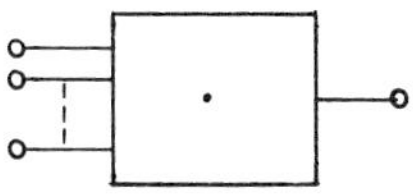

AND — One diode for each input

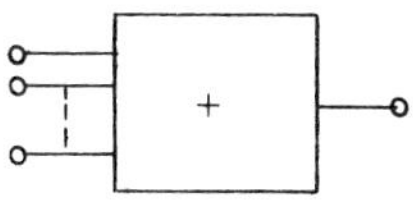

OR — One diode for each input

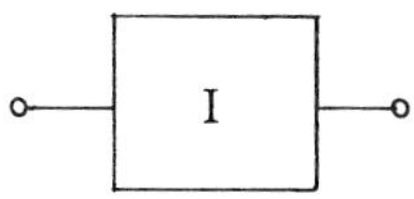

INVERTER — One transistor (or vacuum tube)

Specifying Input–Output Conditions

The job of a logic designer is to design a network of logic blocks which performs some function required in a computer. Each network may have any number of binary inputs, but only one output.

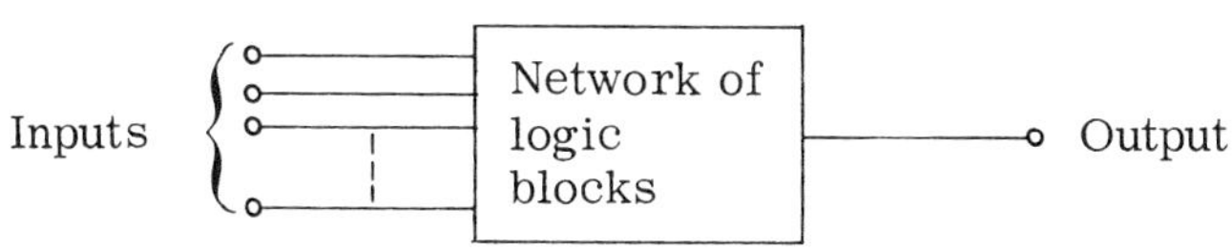

The network may contain any number of logic blocks. A single block may be sufficient, but networks usually contain more than one block.

Generally, only a few types of logic blocks are used in any one computer. The same philosophy has been adopted in this book, and only the three blocks which were introduced earlier are used. These are the AND, OR, and INVERTER blocks.

In the following discussion, unless otherwise stated, we will use the term "circuit" in place of "network of logic blocks." In effect, we are using the term "circuit" in a more general sense than previously.

The circuit designed by a logic designer must satisfy a given set of input-output conditions. In other words, the designer starts to design a circuit knowing only what its output must be in relation to its inputs. The question to be considered now is how these input-output conditions are specified.

Basically, input-output conditions are specified by giving the output level for each possible combination of input levels. This can be done in tabular form as seen in the following example:

Input A	Input B	Input C	Output
0	0	0	1
0	0	1	0
0	1	0	0
0	1	1	0
1	0	0	1
1	0	1	1
1	1	0	1
1	1	1	1

Each row in the table gives one combination of input levels, plus the required output level for that combination of input levels. The first row, for example, says that when all three inputs are at the 0 level, the output must be at the 1 level. The second row says that when inputs A and B are at the 0 level and input C is at the 1 level, the output must be at the 0 level. With this kind of table, the logic designer knows exactly what the output of the circuit must be at all times.

Input-output conditions do not have to be given in tabular form. These conditions are very often expressed by a word statement such as: "The output should be 1 when input A is 1, or when inputs A, B, and C are all 0. The output should be 0 at all other times." Although it seems very different, this word statement contains exactly the same information as the table did. Comparing the word statement with the table, notice that input A is 1 in the last four rows of the table, and inputs A, B, and C all are 0 in the first row of the table, and these are the only rows where the output level is 1. Since they both contain the same information, we could have started with the word statement and converted it into a table, or vice versa.

Word statements commonly give only the input conditions for which the output is at the 1 level. This can be done only if it is assumed that the output is at the 0 level for all other conditions. For example, when we say that the output is at the 1 level if input A and input B are at the 1 level, we still do not know what the output should be for the other input combinations, unless we assume, as we do in this book, that the output is at the 0 level for all other input combinations.

Another comment about word statements; the statement "*A*, *B*, or *C*" means "*A* or *B* or *C*," just as "*A*, *B*, and *C*" means "*A* and *B* and *C*." In addition, when we say "*A* or *B*," we mean "*A* or *B* *or both*." Likewise, "*A*, *B*, or *C*" means "*A* or *B* or *C* alone, or any two (*A* and *B*, or *A* and *C*, or *B* and *C*), or all three (*A* and *B* and *C*)." We call this an *inclusive or* because it includes these other cases.

In addition to tables and word statements, this book will introduce other methods of specifying the input-output conditions of a circuit. The important thing to remember is that every method must, in some way, tell what the output level should be for every possible combination of input levels.

Designing a Circuit

In this section, we will give only a brief introduction to the design of a circuit. Later chapters will cover the subject in more detail and with better methods.

The job of designing a circuit to satisfy a given set of input-output conditions consists of placing the conditions into a form which describes the output in terms of *and*'s, *or*'s and *inverse*'s of the inputs. This is because we are using AND, OR, and INVERTER logic blocks. Let us see how we can do this with the following table:

Input *A*	Input *B*	Output
0	0	0
0	1	1
1	0	1
1	1	1

This table says that the output must be at the 1 level for three of the four input combinations. This can be expressed as the output being at the 1 level for the second *or* third *or* fourth input combination (referring to the second, third, and fourth rows). We can now draw a *partial* circuit.

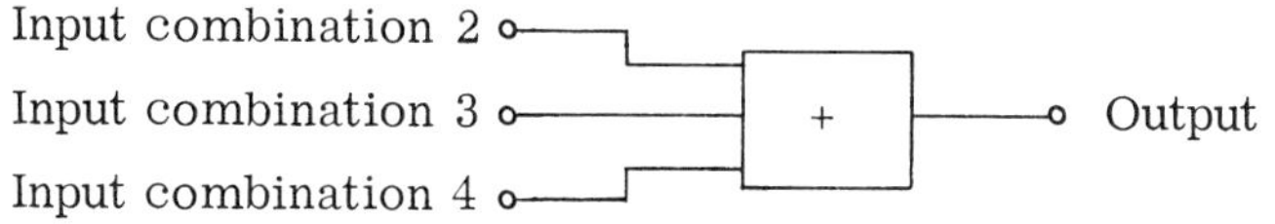

If we assume that the line corresponding to an input combination is at the 1 level when that input combination occurs, then the output will be at the 1 level when the second *or* the third *or* the fourth input combination occurs.

We now need the three lines that correspond to these three input combinations. Considering the second input combination, this line will be at the 1 level when input *A* is at the 0 level *and* input *B* is at the 1 level. We cannot simply use an AND block for this because the output of an AND block is at the 1 level when *all* its inputs are at the 1 level. An AND block can be used if we say that the line should be at the 1 level when the *inverse* of input *A* is at the 1 level *and* input *B* is at the 1 level. It should be remembered that when input *A* is at the 0 level, the *inverse* of input *A* is at the 1 level. Adding this to the partial circuit, we have:

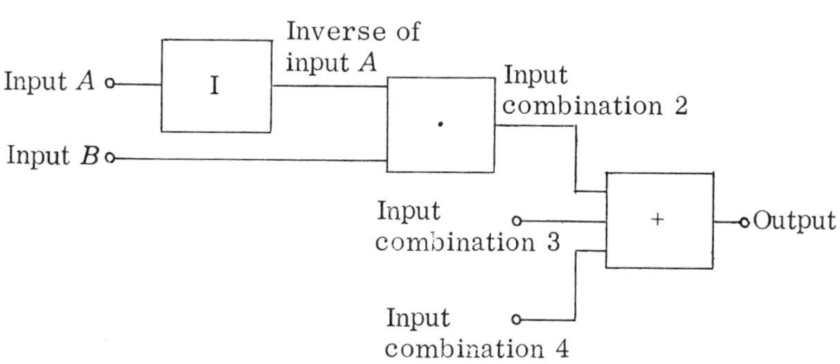

The output of the AND block is at the 1 level when input *A* is at the 0 level and input *B* is at the 1 level. This corresponds to the second input combination.

Using this same reasoning, we can restate the input-output conditions given in the table as: "The output should be at the 1 level when the *inverse* of input *A* is at the 1 level *and* input *B* is at the 1 level, *or* when input *A* is at the 1 level *and* the *inverse* of input *B* is at the 1 level, *or* when input *A* is at the 1 level *and* input *B* is at the 1 level." It is implied that for all other conditions (input combination 1 in this example) the output should be at the 0 level.

We can now draw the complete circuit corresponding to this statement.

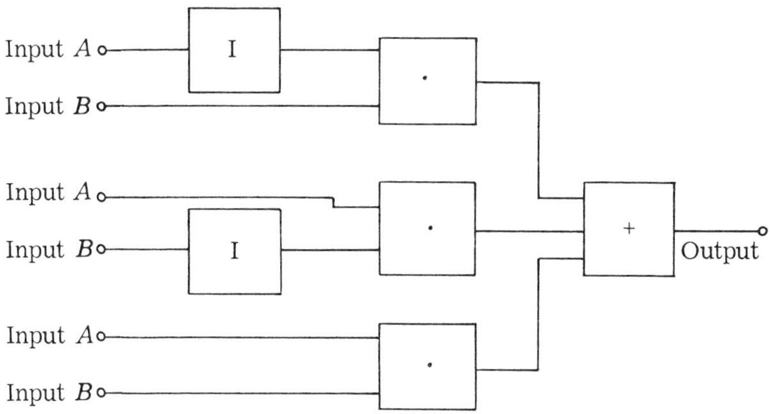

(All lines labeled Input *A* should be connected together, and all lines labeled Input *B* should be connected together. They are shown separated to simplify the drawing.)

If the first combination of inputs occurs, *none* of the AND block outputs will be at the 1 level, and so the circuit output will be at the 0 level. For each of the other three input combinations, one of the AND block outputs will be at the 1 level, and so the circuit output will be at the 1 level.

What we have done is to express a set of input-output conditions in terms of and's, or's, and inverse's of the inputs, and then we drew a circuit to correspond to this statement. If the input-output conditions were *originally* expressed as a word statement in terms of the and's, or's, and inverse's of the inputs, then a circuit could have been designed directly from the statement.

Alternate Circuit Designs

In general, there are many different ways by which a single set of input-output conditions can be expressed in terms of and's, or's, and inverse's. The previous example showed only one of the possible ways to express the set of input-output conditions given in the table. Let us now look at a different way, using the same table, which is repeated here for convenience.

Input A	Input B	Output
0	0	0
0	1	1
1	0	1
1	1	1

Observe that in all rows where either input A or input B (or both) are at the 1 level, the output is also at the 1 level. We can therefore state that: "The output is at the 1 level if input A or input B is at the 1 level." This statement, as usual, implies that the output is at the 0 level for all other conditions, and since the output *is* at the 0 level for the only other input combination, then the statement is valid. We can now draw a circuit to correspond to this new statement.

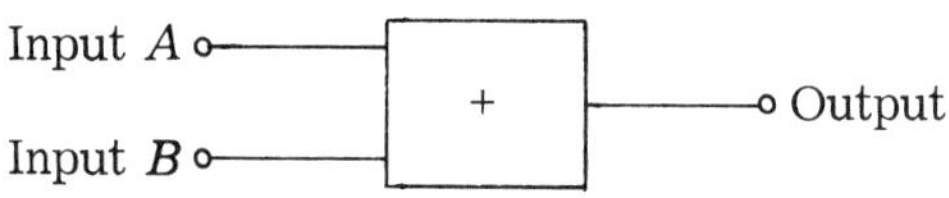

This circuit, which happens to contain only one logic block, is clearly less expensive than the previous circuit, even though the input-output conditions of the two circuits are identical.

The Best Circuit for the Job

We have now reached the point where we can see that the real problem of the logic designer is to find *the least expensive circuit* which satisfies a given set of input-output conditions.

One of the most important tools for helping the logic designer in this task is ***Boolean algebra.*** Boolean algebra enables a circuit to be described by an algebraic equation, which can then be simplified using the theorems of the algebra.

While the theorems of Boolean algebra help to simplify a circuit, they do not always result in the least expensive circuit. However, there are two methods available to the logic designer which will lead him directly to the least expensive circuit. These are called the *Quine-McClusky* method and the *Karnaugh map* method.

Summary

1. All circuits (networks of logic blocks) will be designed using three types of logic blocks: the AND block, the OR block, and the INVERTER block.

2. A circuit is designed to satisfy a given set of input-output conditions. These conditions can be given in tabular form, or as a word statement. Other methods will be introduced in later chapters. No matter in what form they are given, the input-output conditions must supply the output level for every combination of input levels.

3. To design a circuit, the input-output conditions must be expressed in terms of *and*'s, *or*'s, and *inverse*'s of the inputs. A circuit can then be drawn to correspond to this statement.

4. There are many ways to express a set of input-output conditions in terms of *and*'s, *or*'s, and *inverse*'s of the inputs, and Boolean algebra, the Quine-McClusky method, and the Karnaugh map are used to find the one which results in the least expensive circuit.

Chapter 2

BASIC OPERATIONS

This chapter introduces the basic concepts of Boolean algebra by defining its constants, variables, and operations. After defining INVERSION, AND, and OR, these operations are used to combine (or "operate" on) Boolean constants and variables. It is then shown how Boolean algebra can be used as a simple and precise way to describe a logic circuit and how a logic circuit can be designed if its algebraic description is known.

1 **Constant**

A constant is something (value, quantity, etc.) which has a fixed meaning. For example, the constant 5 always means the same thing. In ordinary algebra, there are many possible constants, including all integers and fractions. In Boolean algebra there are only two possible constants. The symbols used for these two constants are 0 and 1. These two constants will be used to describe the two allowable voltage levels on a wire. The Boolean constant 1 will stand for the more positive voltage level (which was called the 1 level), and the Boolean constant 0 will stand for the more negative voltage level (which was called the 0 level).

Variable

A variable is a quantity which can change by taking on the value of any constant in the system. At any one time a variable has the value of a constant, but at a later time it may take on the value of some other constant. As there are only two constants in the system, a variable can take on only one of two values. A variable can be either 0 or 1. Variables are denoted by letters.

Operations

Operations are methods of combining or transforming quantities. In ordinary algebra, the operations of addition, subtraction, multiplication, and division are used to combine quantities. Operations can also be applied to a single quantity, for example, changing the sign of a number. The operations in Boolean algebra are different from the operations in ordinary algebra. The basic operations to be introduced here are INVERSION, AND, and OR. In our application of Boolean algebra to logic design, these basic operations will be used to describe the action of the INVERTER block, the AND block, and the OR block.

2Q The first operation to be defined is INVERSION. This operation operates on a single variable or constant somewhat like the sign in conventional algebra. (Examples of signs are -5, $+A$, $-C$.) To indicate the operation of INVERSION, a bar is placed over the constant or variable to be INVERTED. This operation is defined as follows:

$\overline{1} = 0 \qquad \overline{0} = 1$

When a constant is inverted, it is made equal to the other constant. A 1 with a bar over it is equal to the constant ______.

2A 0

According to the definition, $\overline{1} = 0$.

3Q Inverting the constant 0 makes it equal to ______.

3A 1

$\overline{0} = 1$

4Q Let us now consider the effect of INVERSION on variables. A bar is placed over a variable which is to be inverted. The inverted form of the variable A is ______.

4A $\overline{A}$

A bar is placed over the variable A to change it to its inverted form.

5Q A variable without a bar over it is said to be in its true form. The true form of the variable A is written as ______.

5A A

The true form of the variable has no bar over it. The inverted form of the variable has a bar over it.

6Q The two possible values of the variable A are 0 and 1. If, at a particular time, $A = 1$, then at that time $\overline{A} =$ ______.

6A $\overline{1}$ or 0

If $A = 1$, then $\overline{A} = \overline{1}$ and $\overline{1} = 0$.

7Q The Boolean operation of INVERSION can be used to describe the operation of the INVERTER circuit, which was introduced in Chap. 1.

If the input to an INVERTER is given as a Boolean constant instead of a voltage, the output of the inverter is the inverse of that constant. When the input to an INVERTER is 1, the output is _____.

7A $\overline{1}$ or 0

The output is the *inverse* of the input, 1 to 0.

8Q When the input to the INVERTER is 0, the output is _____.

8A $\overline{0}$ or 1

9Q We may use variables to summarize the relation of input to output of an INVERTER; thus if the input is C, the output is _____.

9A $\overline{C}$

When the value of C is 1, the value of $\overline{C}$ is 0, but the output of the INVERTER is also equal to 0. When the value of C is 0, the value of $\overline{C}$ is 1, but the output of the INVERTER is also equal to 1.

10 **Summary**

1. $\overline{1} = 0$; $\overline{0} = 1$

2. When $A = 0$, then $\overline{A} = 1$; when $A = 1$, then $\overline{A} = 0$

3. The two forms of a variable are the *true* form and the *inverted* form.

4. If the input signal of an inverter is assigned a variable which is in true form, then the output signal is represented by the same variable but in inverted form.

---

11 The next operation to be introduced is the AND, which connects two or more constants and/or variables. The dot (·), which is used for multiplication in conventional algebra, is used in Boolean algebra as the symbol for the AND operation. The AND operation is defined as follows:

$0 \cdot 0 = 0 \qquad 1 \cdot 0 = 0$
$0 \cdot 1 = 0 \qquad 1 \cdot 1 = 1$

Using the similarity with the English conjunction "and," the following mental aid can be used: The AND of two constants is 1 only if the first *and* the second are 1. (*Note*: To distinguish between the Boolean operation and the English conjunction, the Boolean operation is always written in capital letters.)

12Q $1 \cdot 1 =$ ______

12A 1

According to the definition, $1 \cdot 1 = 1$.

13Q $1 \cdot 0 =$ ______

13A 0

14Q $0 \cdot 0 =$ ______

14A 0

15Q The AND operation was defined for two constants ANDed (connected by the operation AND) together, but three or more constants can be ANDed together by the following procedure: First AND any two constants together, then AND the result of that operation with a third constant, and so on. Parentheses have the same meaning in Boolean algebra as they do in conventional algebra. In both, the operation within the parentheses is to be performed before the operation outside the parentheses. Performing the operation within the parentheses first, find the constant for the following:

$$1 \cdot (1 \cdot 1) = \_\_\_\_\_\_$$

15A 1

In the parentheses the $1 \cdot 1$ gives 1, thus the 1 outside is ANDed with 1 to give the answer of 1. Pictorially:

$$1 \cdot \underbrace{(1 \cdot 1)}$$

$$\underbrace{1 \quad \cdot \quad 1}$$

$$1$$

16Q Evaluate the following:

$$0 \cdot (1 \cdot 1) = \_\_\_\_\_\_$$

16A 0

$$0 \cdot \underbrace{(1 \cdot 1)}$$

$$\underbrace{0 \quad \cdot \quad 1}$$

$$0$$

---

17 In the last few frames the AND operation was extended by the use of parentheses. It turns out that these parentheses are not needed because they are redundant. An expanded definition is required.

Definition

If two or more constants are ANDed together, all have to be 1 for the result to be 1.

The mental aid is now: The AND of several constants is 1 if the first constant, *and* the second constant, . . . , *and* the last constant is 1.

18Q $0 \cdot 1 \cdot 1 \cdot 1 \cdot 0 =$ ______

18A 0

Because not all the constants are 1's.

19Q Using the information learned thus far, you should be able to answer this question: If $A \cdot B$ is equal to 1, then $A =$ ______ and $B =$ ______ .

19A $A = 1$; $B = 1$

If either of them were 0, there would be one 0 ANDed with one 1 and the answer would be 0.

20Q We have introduced the AND operation in Boolean algebra, but what is the relationship between this operation and the AND block that was described previously? It turns out that the output of the AND block may be expressed as the algebraic AND of the inputs. For example,

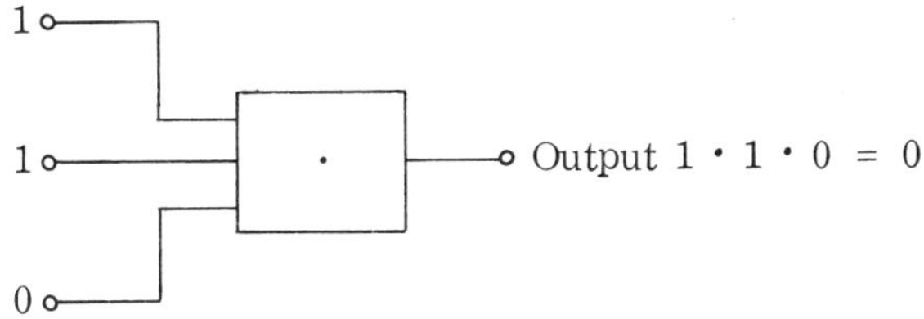

What is the output of the following circuit?

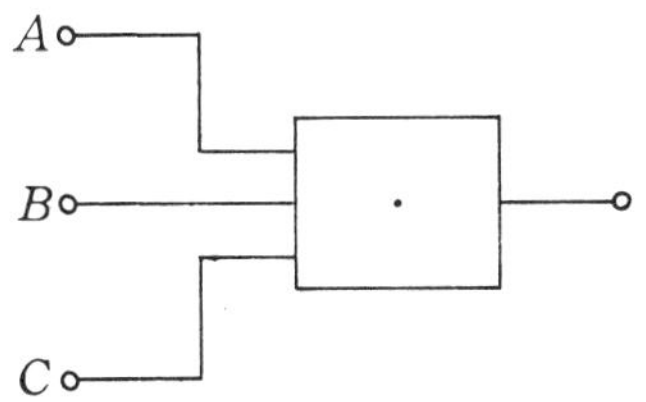

20A $A \cdot B \cdot C$

The output is the AND of the inputs A, B, and C.

21Q When a constant (or variable) which is to be inverted is ANDed with other constants (or variables), the INVERSION is performed first. After the result of the inversion has been found, this result is ANDed with the other constants (or variables):

$$1 \cdot 1 \cdot \overline{1} \cdot 1 \cdot 1 = \_\_\_\_\_$$

21A 0

$$1 \cdot 1 \cdot \underbrace{\overline{1}} \cdot 1 \cdot 1$$

$$\underbrace{1 \cdot 1 \cdot 0 \cdot 1 \cdot 1}_{0}$$

The $\overline{1}$ is replaced by a 0. This 0 is ANDed with the other 1's, making the overall value equal to 0.

22Q $1 \cdot 1 \cdot \overline{0} \cdot 1 \cdot \overline{0} = \_\_\_\_\_$

22A 1

$$1 \cdot 1 \cdot \underbrace{\overline{0}} \cdot 1 \cdot \underbrace{\overline{0}}$$

$$\underbrace{1 \cdot 1 \cdot 1 \cdot 1 \cdot 1}_{1}$$

The inverted 0's are replaced by 1's, and since all the constants are 1's, the result is 1.

23Q If $A \cdot \overline{0} \cdot \overline{B} \cdot 1 = 1$, then $A =$ _____ and $B =$ _____.

23A A = 1; B = 0

The values of A and $\overline{B}$ must be 1 to make $A \cdot 1 \cdot \overline{B} \cdot \overline{0} = 1$. If $\overline{B} = 1$ it follows that $B = 0$.

24Q We have already described the input-output relations of the INVERTER and the AND blocks using the Boolean operations of INVERT and AND. Let us now try to describe the input-output relation of a circuit consisting of two logic blocks. What is the output expressed in terms of the inputs of the following circuit?

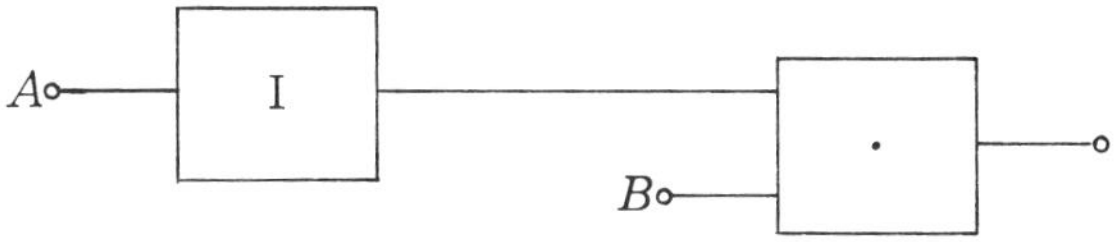

24A $\bar{A} \cdot B$

The output of the INVERTER is $\bar{A}$, and this is ANDed with B to form $\bar{A} \cdot B$.

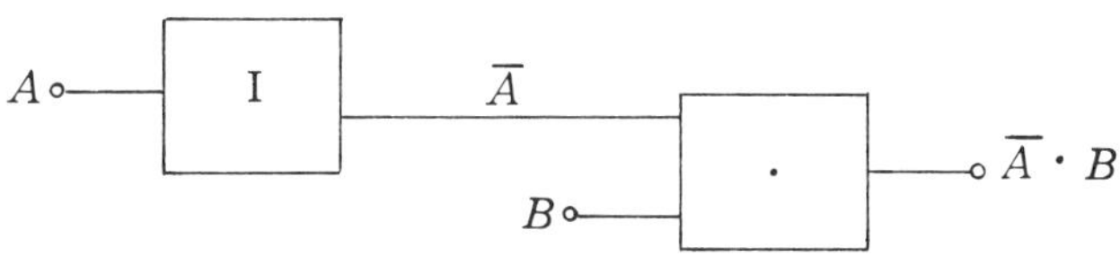

25Q In the previous example we determined the output of a given circuit. It is also possible to obtain the circuit if the required output is given. If we must design a circuit whose output is $\bar{A} \cdot B$, we have to have an AND circuit with $\bar{A}$ and B as inputs. If A is available but not $\bar{A}$, we must obtain $\bar{A}$ from A by using an INVERTER. Draw the circuit whose output is $\bar{A} \cdot \bar{B}$, assuming that only A and B are available as inputs.

25A 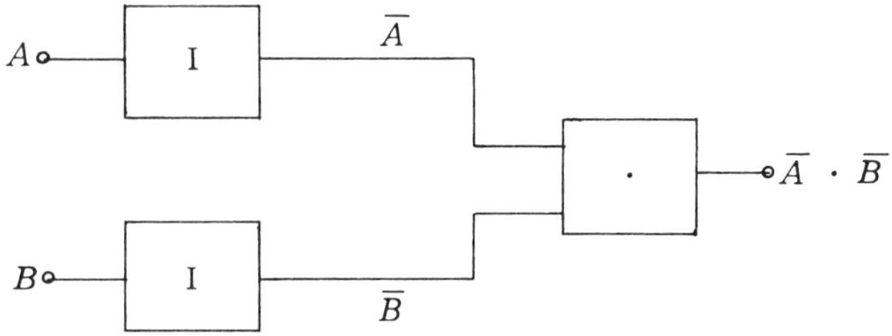

26 There are several additional points of interest about the AND operation:

1. In conventional algebra the multiplication sign is often left out; in Boolean algebra it is also customary to omit the AND sign:

$$A \cdot B \cdot C = ABC$$

2. The AND is commutative. This means that the order in which the variables (or constants) are written will not affect the result:

$$AB = BA$$

3. The AND is associative. This means that the order in which the variables or constants are actually ANDed does not change the result. For example, if we were to AND only pairs of variables, the following equation would hold:

$$A \cdot (B \cdot C) = (A \cdot B) \cdot C = (A \cdot C) \cdot B$$

All these properties are not new; the operations of addition and multiplication in conventional algebra possess them too.

27 **Summary for AND**

1. $0 \cdot 0 = 0 \qquad 1 \cdot 0 = 0$
   $0 \cdot 1 = 0 \qquad 1 \cdot 1 = 1$

2. If more than two constants are ANDed together, all have to be 1 for the result to be 1. *Mental aid:* The AND of several constants is 1 if the first constant, *and* the second constant, . . . , *and* the last constant is 1.

3. Logic block

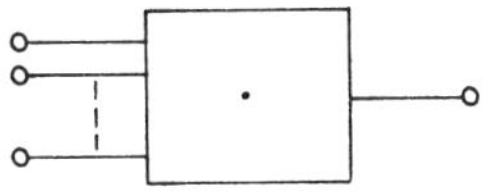

4. When a constant (or variable) which is inverted is ANDed with other constants (or variables), the INVERSION is performed first. This is true when calculating its value *and* when drawing the circuit.

---

28 The last operation to be defined is the OR. The symbol used for the OR is the +. This operation connects two or more constants or variables and is defined as follows:

$0 + 0 = 0 \qquad 1 + 0 = 1$
$0 + 1 = 1 \qquad 1 + 1 = 1$

Using the English conjunction "or," a mental aid can be formed again: The OR of two constants is 1 if the first *or* the second (*or* both) are 1.

29Q $1 + 0 =$ _____

29A 1

30Q 0 + 0 = ______

30A 0

There must be at least one 1 to make the result equal to 1.

31Q 1 + 1 = ______

31A 1

If one input *or* the other input (*or* both) is 1, the result will be 1.

32Q The OR was defined as an operation which connects together two constants. If we want to OR together more than two constants, we can first OR any two together, then OR the result with a third constant, etc. The order in which this is performed can be shown by means of parentheses. What is 1 + (0 + 0) = ______ ?

32A 1

$$1 + \underbrace{(0 + 0)}$$

$$\underbrace{1 \; + \; 0}$$

$$1$$

33Q When ORing together more than two constants (taken two at a time), it turns out that the result is 1 if one or more of the constants is 1. We can use this fact to expand the definition of the OR for any number of constants.

Definition

For any number of constants ORed together, one or more have to equal 1 for the result to equal 1.

The aid is extended too: The result is 1 if the first constant is 1, *or* the second is 1, ..., *or* the last is 1.

$0 + 1 + 0 + 0 + 1 =$ _____

33A 1

If at least one of the constants ORed together is a 1, the answer is 1.

34Q When $A + B + C = 0$, then the variables must have the following values:

$A =$ ________ $B =$ ________ $C =$ ________

34A $A = 0$; $B = 0$; $C = 0$

If any variable is equal to 1, the value of $(A + B + C)$ is 1, and the equality does not hold.

35Q We have found that the output of an AND circuit is the AND operation of all the inputs. The same relationship holds for the OR operation and OR circuit. The output of the OR circuit is the algebraic OR operation of the inputs into the OR circuit. What is the output of the circuit shown?

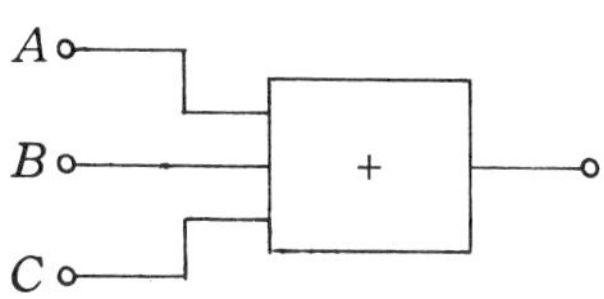

35A $A + B + C$

The output is the OR of the inputs.

36Q When a constant (or variable) which is to be inverted is ORed with other constants (or variables), the inversion is performed first.

$$\overline{1} + 0 + \overline{0} = \_\_\_\_\_\_ \text{ (0 or 1?)}$$

36A 1 $\underbrace{\overline{1}} + 0 + \underbrace{\overline{0}}$

$$\underbrace{0 + 0 + 1}$$

$$1$$

37Q When designing a circuit whose output is $\overline{A} + B + \overline{C} + D$, given inputs in true form only, the logic blocks are placed in the same order (going from input to output) in which the operations would be performed. Sketch the circuit for $\overline{A} + B + \overline{C} + D$.

37A

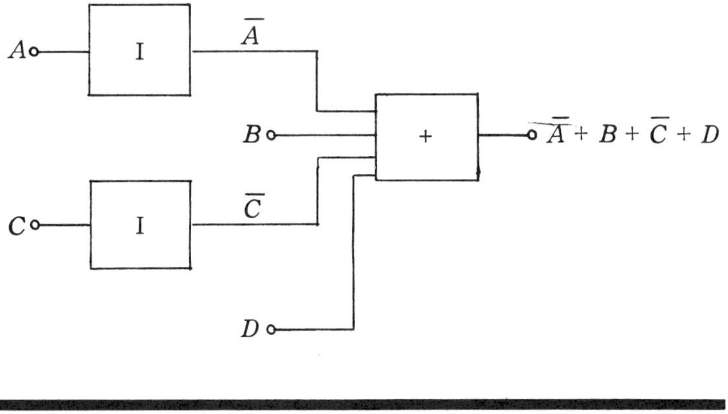

38 Here are two additional properties of the OR operation:

1. OR is commutative:

$$A + B + C = A + C + B = B + A + C$$

The order does not change the result.

2. OR is associative:

$$A + B + C = (A + B) + C = A + (B + C)$$

If the variables were ORed in pairs, the result would not change.

39 **Summary**

AND

$$0 \cdot 0 = 0$$
$$0 \cdot 1 = 0$$
$$1 \cdot 0 = 0$$
$$1 \cdot 1 = 1$$

If two or more constants and/or variables are ANDed, each one has to be 1 for the answer to be 1.

The AND is associative and commutative.

INVERSION is performed before AND.

The logic block implementing the AND operation is:

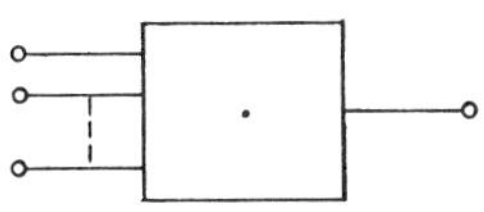

OR

$$0 + 0 = 0$$
$$0 + 1 = 1$$
$$1 + 0 = 1$$
$$1 + 1 = 1$$

If two or more constants and/or variables are ORed, at least one has to be 1 for the answer to be 1.

The OR is associative and commutative.

INVERSION is performed before OR.

The logic block implementing the OR operation is:

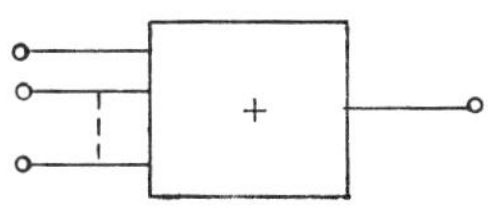

PROBLEMS

1. If $A \cdot \overline{B} \cdot C = 1$, what value must A, B, and $\overline{C}$ have?

2. Design a circuit whose output is $\overline{A} + \overline{B}$ using inputs A and $\overline{B}$.

3. Design a circuit whose output is $\overline{A} \cdot \overline{B} \cdot \overline{C}$ using inputs A, $\overline{B}$, and C.

4. In practice, the situation arises when either the true form of a variable or the inverted form, or both, are given. If only one form is available, the other form may have to be generated. If both forms are available, the proper one can be selected; inverters are never needed. Design a circuit whose output is $A + \overline{B} + \overline{C}$. The following inputs are available: A, B, C, $\overline{C}$.

5. What value must A and B have to make the outputs of both circuits be 1 at the same time?

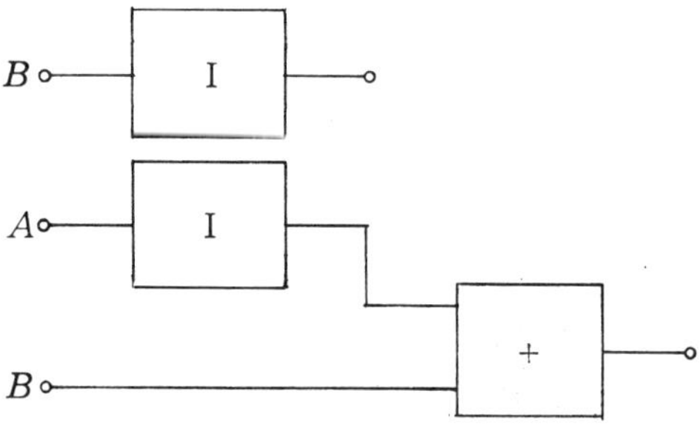

6. What value of A and B, if any, will cause both outputs to have the same value?

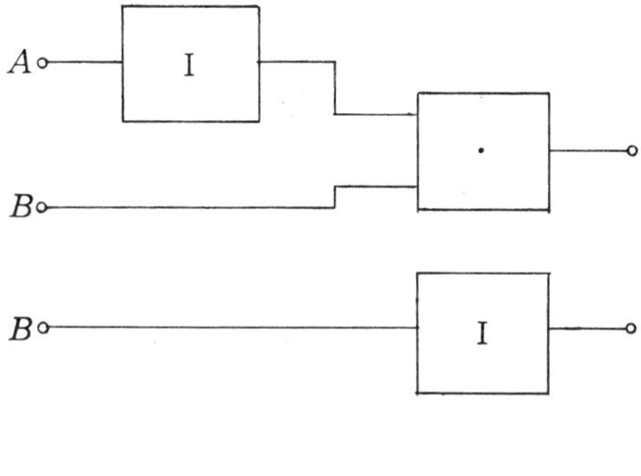

Chapter 3

TRUTH TABLE AND BASIC THEOREMS

This chapter introduces the truth table, which contains a list of the values of a Boolean expression for each possible combination of values of the variables in the expression. After explaining the construction of truth tables for any number of variables, it is shown how an expression may be plotted, or entered, in the table. The first application of the truth table is in determining whether two expressions are equivalent. A second application is in proving the validity of simple one-variable theorems.

---

1 Before starting the discussion of the truth table the word "expression" will be defined.

Definition

An *expression* is a string of constants and/or variables, in true or inverted form, connected by one or more operations. It may include parentheses. If a string contains an equality sign (=), the string is *not* an expression. To repeat, an expression cannot contain an equality sign. Examples of expressions are

$A + 1$ $\quad$ $\overline{B} + \overline{D} \cdot E$

$C + D$ $\quad$ $(A + \overline{B}) \cdot (C + \overline{D})$

2Q Which of the following are not expressions?

1. $AB + \overline{C}$
2. $\overline{A} + \overline{A}B = \overline{A}$
3. $\overline{A} = \overline{A}$
4. $\overline{A} + \overline{A} + \overline{B}$

2A 2 and 3

Both contain an equality sign.

3Q If two expressions are ORed together, is the result an expression?

3A Yes

The result still does not contain an equality sign.

4 The name *truth table* comes from a similar table used in symbolic logic, in which the "truth" or "falsity" of a statement is listed for all possible conditions. The truth table has two parts. The left part of the table contains all combinations of values of the variables used in an expression, and the right part contains the value of the expression for each combination of values listed on the left side. We will start out by developing the left part of the table.

5Q How many values can the Boolean variable C take on?

5A Two

The variable must be either 1 or 0; there is no other value it can take on.

6Q Let us put the values of C in tabular form. This table is the left half of a truth table for an expression which contains only one variable, C:

C
0
1

These are all possible values of C. Let us add another variable, B. What are the possible combinations of values of the two variables B and C? (B can be either 0 or 1; C can be either 0 or 1.)

6A These are four possible combinations:

$B = 0$ and $C = 0$	$B = 1$ and $C = 0$
$B = 0$ and $C = 1$	$B = 1$ and $C = 1$

7Q From this result we can see that, for the two values of C, the value of B can be 0. For the same two values of C, the value of B can also be 1. In tabular form this is written as (fill in the blanks):

B	C
___	0
___	1
___	0
___	1

7A

B	C
0	0
0	1
1	0
1	1

In the first two rows, B is 0 when C is both 0 and 1. In the last two rows, B is 1 when C is both 0 and 1.

---

8 **Summary**

The table you note above is the left half of a two-variable truth table which was derived from the one-variable table. Any table can be extended by one variable by these steps:

1. Place the new variable to the left of the others and enter a 0 in all rows under the new variable.
2. Repeat the original table below the old one.
3. Place a 1, under the new variable, in each row added in step 2.

9Q Using the three steps described in the last frame, derive the left half of a two-variable truth table from the one-variable truth table by adding the variable *B* to the table for *C*:

C
0
1

9A The two-variable table is:

B	*C*
0	0
0	1
1	0
1	1

The first step is to place *B* to the left of *C* and add two 0's under *B*. The table takes on the form shown on the right. The second step is to repeat the original table. This means that column *C* will be extended by a 0 and a 1. To get the final table add two 1's under the 0's in column *B*.

B	*C*
0	0
0	1

10Q The two-variable table has four rows. Are there any other combinations of values of B and C which would yield a new row not yet in the table?

10A No

The two-variable truth table had been derived from the one-variable truth table by permitting the new variable to take on both 0 and 1 for every possible value of the original truth table. Thus if the original table included all possible values for one variable, the new table includes all possible combinations of values for two variables.

11Q Derive a three-variable table from the two-variable table by adding A.

B	C
0	0
0	1
1	0
1	1

11A

A	B	C
0	0	0
0	0	1
0	1	0
0	1	1
1	0	0
1	0	1
1	1	0
1	1	1

12 By using the three simple steps shown earlier, the left half of a truth table of any size can be derived by adding one variable at a time. The computation for tables of various sizes is shown below. The number of rows in the table with n variables is twice the number of rows in the table with $(n - 1)$ variables.

$2 \times 2^0 = 2^1 = 2$	(rows in a one-variable table)
$2 \times 2^1 = 2^2 = 4$	(rows in a two-variable table)
$2 \times 2^2 = 2^3 = 8$	(rows in a three-variable table)
$2 \times 2^3 = 2^4 = 16$	(rows in a four-variable table)

Note that the number of rows grows rapidly; for example, a 10-variable truth table has 1,024 rows ($1{,}024 = 2^{10}$).

13Q Thus far only half of the truth table has been shown. To complete the table a new column is added, separated from the left part of the table by a vertical line.

A	B	$A + \overline{B}$
0	0	1
0	1	0
1	0	1
1	1	1

The expression for which the table is written heads the new column. Under it are written the values of the expression for all the combinations of values of the variables. For example, the first row in the table states that for $A = 0$, $B = 0$, the value of the expression $A + \overline{B}$ is 1. (Is this correct?) The second row in the table states that for $A =$ _____, $B =$ _____, the value of the expression $A + \overline{B}$ is _____.

13A $A = 0$; $B = 1$; $A + \overline{B} = 0$

If $B = 1$, $\overline{B} = 0$; thus $A + \overline{B} = 0 + 0 = 0$

14Q To find the value of the expression for any row, replace each variable in the expression by the value of the variable given on the left side of the table. For example, if $A = 0$ and $B = 0$, the expression $A + B$ is equal to $0 + 0$ which equals 0. Thus the entry in the first row is 0.

A	B	$A + B$
0	0	0
0	1	
1	0	
1	1	

In the second row $A = 0$, $B =$ _____, and $A + B =$ _____ + _____ = _____.

14A $B = 1$; $A + B = 0 + 1 = 1$

15Q

A	B	$A + B$
0	0	0
0	1	1
1	0	
1	1	

The remaining two entries in the table are _____ and _____.

15A 1 and ~~1~~

For $A = 1$ and $B = 0$, the expression is $1 + 0$, which has the value 1. For the last row, $1 + 1 = 1$.

16 The entire table for the expression $A + B$ is:

A	B	$A + B$
0	0	0
0	1	1
1	0	1
1	1	1

17Q The table for the expression $A \cdot B$ is started below. Complete the table by filling in the last column.

A	B	$A \cdot B$
0	0	
0	1	
1	0	
1	1	

17A

A	B	$A \cdot B$
0	0	0
0	1	0
1	0	0
1	1	1

The first three entries in the last column are 0 because at least one of the variables is equal to 0, which is enough to make the entire expression equal to 0. In the last row, $A = 1$ and $B = 1$; thus $A \cdot B = 1 \cdot 1 = 1$.

18Q Sometimes more complicated expressions are plotted on the truth table. Usually they can be plotted more easily by first plotting partial answers or intermediate steps. For example, when plotting the expression $\overline{A} + B$, the inverse of A, $\overline{A}$ can be plotted first. Fill in the column for $\overline{A}$. In this example a one-variable expression, namely $\overline{A}$, is plotted on a two-variable table. *Note: If the inverse form of a variable is used, the table is still written for the true form of the variable.*

A	B	$\overline{A}$
0	0	
0	1	
1	0	
1	1	

18A

A	B	$\overline{A}$
0	0	1
0	1	1
1	0	0
1	1	0

Whenever $A = 0$, $\overline{A} = 1$; whenever $A = 1$, $\overline{A} = 0$.

19Q Having both $\overline{A}$ and B available on the table, the OR of these two columns will produce the desired answer. Note that more than one expression can be plotted on one table. To separate them vertical lines are placed between them. Complete the table.

A	B	$\overline{A}$	$\overline{A} + B$
0	0	1	
0	1	1	
1	0	0	
1	1	0	

19A

A	B	$\overline{A}$	$\overline{A} + B$
0	0	1	1
0	1	1	1
1	0	0	0
1	1	0	1

20Q Try the expression $\overline{A} \cdot \overline{B}$. Here you might want to take two intermediate steps, $\overline{A}$ and $\overline{B}$.

A	B	$\overline{A}$	$\overline{B}$	$\overline{A} \cdot \overline{B}$
0	0			
0	1			
1	0			
1	1			

20A

A	B	$\overline{A}$	$\overline{B}$	$\overline{A} \cdot \overline{B}$
0	0	1	1	1
0	1	1	0	0
1	0	0	1	0
1	1	0	0	0

If either $\overline{A}$ or $\overline{B}$ is 0, then $\overline{A} \cdot \overline{B}$ is equal to 0.

21Q Below is a three-variable expression on a three-variable table. Fill in *all* the entries in this table.

A	B	C	$\overline{A}$	$\overline{C}$	$\overline{A} + B + \overline{C}$
0	0	0			
0	0	1			
0	1	0			
0	1	1			
1	0	0			
1	0	1			
1	1	0			
1	1	1			

21A

A	B	C	$\overline{A}$	$\overline{C}$	$\overline{A} + B + \overline{C}$	
0	0	0	1	1	1	(1 + 0 + 1)
0	0	1	1	0	1	(1 + 0 + 0)
0	1	0	1	1	1	(1 + 1 + 1)
0	1	1	1	0	1	(1 + 1 + 0)
1	0	0	0	1	1	(0 + 0 + 1)
1	0	1	0	0	0	(0 + 0 + 0)
1	1	0	0	1	1	(0 + 1 + 1)
1	1	1	0	0	1	(0 + 1 + 0)

This problem is a direct extension of previous ones. To aid you, the expressions that were evaluated to obtain the results are written on the right. They are *not part of the truth table.*

22 You are now ready for the second half of the chapter. The truth table will be used as a tool to determine whether two expressions are equivalent. The check consists of comparing the values of the two expressions for *all* possible combinations of the values of the variables.

23Q To determine whether two expressions are equivalent, we can use a truth table consisting of three parts. The first part contains variables in the expressions, the second part contains one of the expressions, and the third part contains the other expression. If the values of the two expressions are the same in every row, the expressions are equivalent. Let us determine if the expression $A + 0$ is equivalent to the expression A. Fill in the last two columns.

Variables A	First expression $A + 0$	Second expression A
0		
1		

23A

A	$A + 0$	A	
0	0	0	$(0 + 0 = 0)$
1	1	1	$(1 + 0 = 1)$

24 Since the value of the expression $A + 0$ is equal to the value of the expression A for all possible cases, we can state the following theorem:

Theorem

$$A + 0 = A$$

25Q Let us try to prove the theorem $A + 1 = 1$. Complete the table and determine whether $A + 1 = 1$.

Variables A	First expression $A + 1$	Second expression 1
0		
1		

25A Yes, they are equal.

A	$A + 1$	1
0	1	1
1	1	1

$A + 1 = 1$, since their values are the same in all rows. Note that the first expression is independent of the value of A.

26 **Theorem**

$$A + 1 = 1$$

27Q Prove the theorem $A + A = A$.

27A

A	$A + A$	A	
0	0	0	$(0 + 0 = 0)$
1	1	1	$(1 + 1 = 1)$

28 **Theorem**

$$A + A = A$$

29Q Prove the theorem $A + \overline{A} = 1$.

29A

A	$A + \overline{A}$	1
0	1	1
1	1	1

30 **Theorem**

$$A + \overline{A} = 1$$

Here is a list of the theorems proved thus far:

$$A + 0 = A \qquad A + A = A$$
$$A + 1 = 1 \qquad A + \overline{A} = 1$$

31Q Let us now look at some similar theorems using the AND operation. This time, however, you will be asked to complete the theorem yourself by finding the simplest expression which is equivalent to the given expression.

$A \cdot 0 =$ ____

A	$A \cdot 0$	?
0	0	0
1	0	0

31A $A \cdot 0 = 0$

A	$A \cdot 0$	0
0	0	0
1	0	0

32 **Theorem**

$$A \cdot 0 = 0$$

33Q $A \cdot 1 =$ ______

A	$A \cdot 1$	?
0	0	0
1	1	1

33A $A \cdot 1 = A$

A	$A \cdot 1$	A
0	0	0
1	1	1

34 **Theorem**

$$A \cdot 1 = A$$

35Q $A \cdot A =$ ______

35A $A \cdot A = A$

A	$A \cdot A$	A
0	0	0
1	1	1

36 **Theorem**

$$A \cdot A = A$$

37Q $A \cdot \overline{A} =$ ______

37A $A \cdot \overline{A} = 0$

A	$A \cdot \overline{A}$	0
0	0	0
1	0	0

38 **Theorem**

$$A \cdot \overline{A} = 0$$

Summary

Here is a list of all the theorems which have been proved thus far.

$A + 0 = A$	$A + A = A$	$A \cdot 0 = 0$	$A \cdot A = A$
$A + 1 = 1$	$A + \overline{A} = 1$	$A \cdot 1 = A$	$A \cdot \overline{A} = 0$

39Q All these theorems have been written in terms of the variable A, but they hold true for all variables. The variable A can also stand for an inverted variable such as $\overline{B}$. Replacing A by $\overline{B}$ in the theorem $A \cdot 0 = 0$, we get $\overline{B} \cdot 0 = 0$. No matter what the value of $\overline{B}$ is, the result of $\overline{B}$ AND 0 is 0. What is $\overline{B} + 0$ equal to?

39A $\overline{B} + 0 = \overline{B}$

$\overline{B}$ has replaced A in the theorem $A + 0 = A$.

40Q The theorem governing the substitution of variables by other variables is called the *substitution theorem*. Using the substitution theorem, find another theorem from the list in frame 38 which is equivalent to $\overline{A} \cdot 1 = \overline{A}$.

40A Theorem $\overline{A} \cdot 1 = \overline{A}$ is equivalent to $A \cdot 1 = A$ if $\overline{A}$ is replaced by A.

41Q Complete the equation twice to obtain equations equivalent to those in frame 38, page 49:

$B \cdot \_\_\_\_\_ = 0$ $B \cdot \_\_\_\_\_ = 0$

41A $B \cdot \underline{0} = 0$
$B \cdot \overline{B} = 0$

PROBLEMS

1. Write the truth table for the expression $A + \overline{B}$ *without* adding a column for $\overline{B}$.

2. How many rows are needed in a truth table for the expression $A + \overline{A} + B$?

3. Write the truth table for the expression $A \cdot B \cdot \overline{C}$.

4. Use truth tables to prove (*a*) $A + \overline{B} + C + \overline{C} = 1$; (*b*) $A \cdot \overline{B} \cdot C \cdot \overline{C} = 0$.

5. Using the theorems presented in this chapter, prove that (*a*) $A \cdot \overline{B} \cdot 0 = 0$; (*b*) $A + A + \overline{A} + 1 = 1$.

   (*Note:* An expression such as $A \cdot \overline{B} \cdot 0$ may be written as $A \cdot (\overline{B} \cdot 0)$.)

6. Find the outputs of the following circuits by first finding the output of each block, going from left to right, and simplifying as you go along.

   (*a*)

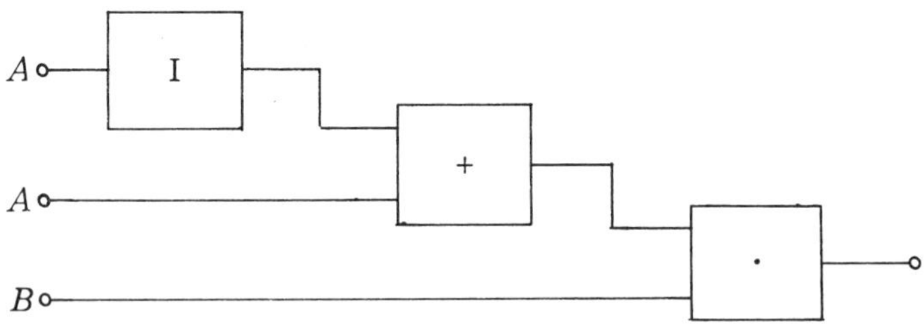

(b)

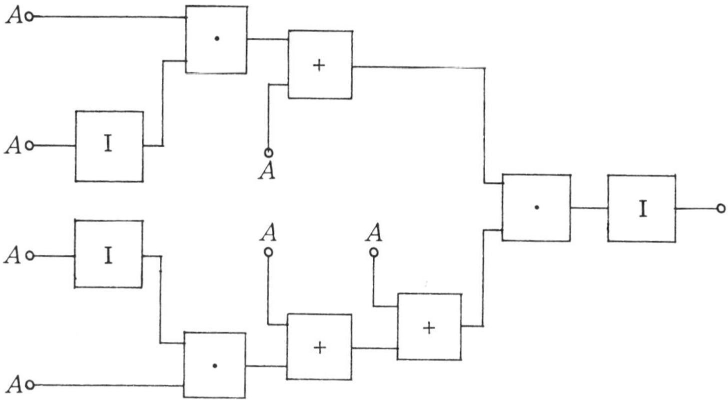

Chapter 4

ADVANCED THEOREMS

This chapter continues theorem proving with theorems which contain more than one variable and which include both the AND and the OR operations. The proved theorems are then used to simplify expressions without changing the value of the expressions. The next step is to show how algebraic simplification is used to simplify logic circuits. Using the theorems, various circuits are simplified, and the cost is computed in accordance with set rules.

---

1 Thus far nothing has been stated about the priority of AND over OR or vice versa. In Boolean algebra the assumed order of priority is

1. AND 2. OR

You are already familiar with a similar rule in conventional algebra where multiplication has priority over addition.

2Q It was shown in Chap. 2, page 33, that INVERSION is performed even prior to AND or OR.

$$1 + 0 + \overline{1} \cdot 0$$

In the expression above, the order of operations is:

1. ________ 2. ________ 3. ________

2A The expression is: $1 + 0 + \underbrace{\overline{1}} \cdot 0$

INVERSION: $1 + 0 + \underbrace{0 \cdot 0}$

AND: $\underbrace{1 + 0 + 0}$

OR: 1

(Evaluation was not required.)

3Q Evaluate the following expression by performing the highest-priority operations first and continuing until the expression reduces to a single constant. (See expression evaluated in frame 2A.)

$\overline{1} + \overline{1} \cdot 0 + 0$

3A

$\underbrace{\overline{1}} + \underbrace{\overline{1}} \cdot 0 + 0$ INVERSION

$0 + \underbrace{0 \cdot 0} + 0$ AND

$\underbrace{0 + 0 + 0}$ OR

0 ANSWER

---

4 In Boolean algebra the AND sign is usually omitted. For example:

$A \cdot B \cdot C$	is written as ABC
$A \cdot (B + C)$	is written as $A(B + C)$
$(A + B) \cdot (C + D)$	is written as $(A + B)(C + D)$

5Q We are now ready to start to prove theorems. Let us first prove the theorem $A + AB = A$ by finding how the value of $A + AB$ depends on the values of A and B. On the truth table plot the values of the expression AB.

A	B	AB
0	0	
0	1	
1	0	
1	1	

5A

A	B	AB
0	0	0
0	1	0
1	0	0
1	1	1

6Q In proving the theorem $A + AB = A$, plot the expression $A + AB$.

A	B	AB	$A + AB$
0	0	0	
0	1	0	
1	0	0	
1	1	1	

6A

A	B	AB	$A + AB$
0	0	0	0
0	1	0	0
1	0	0	1
1	1	1	1

7Q Is the statement $A = A + AB$ correct? Why?

7A Yes

Comparing the value for A and the value for $A + AB$, we find the two are the same for each row. Therefore the two expressions have the same value for every combination of variables, and the expressions are equal.

8 **Theorem**

$$A + AB = A$$

9 In the next few frames theorems will be proved. This can be done on the truth table by testing their equality for every possible combination.

1. A truth table is set up for as many variables as appear in the theorem.

2. The expression on the left side of the equation is evaluated. This may be done in several steps.

3. The right side is evaluated. This again may take several steps.

4. The two columns obtained are compared. If they differ in at least one row, the theorem is not correct.

10Q Apply the method for proving the correctness of this equation

$$A + \bar{A}B = A + B$$

The first step is to plot $\bar{A}B$; the second step is to plot $A + \bar{A}B$. (Use the table on page 56.)

A	B	$\bar{A}B$	$A + \bar{A}B$
0	0		
0	1		
1	0		
1	1		

10A

A	B	$\bar{A}B$	$A + \bar{A}B$
0	0	0	0
0	1	1	1
1	0	0	1
1	1	0	1

11Q The right-hand side of the equation is evaluated next.

A	B	$A + B$
0	0	
0	1	
1	0	
1	1	

11A

A	B	$A + B$
0	0	0
0	1	1
1	0	1
1	1	1

12Q The last step is to compare the results. Are the two expressions $A + \bar{A}B$ and $A + B$ equal?

A	B	$A + \bar{A}B$	$A + B$
0	0	0	0
0	1	1	1
1	0	1	1
1	1	1	1

12A Yes

All corresponding entries are alike.

13 **Theorem**

$$A + \overline{A}B = A + B$$

14Q Lastly let us investigate $A(A + B) = A$. (*Note:* First A and B are ORed, then the result is ANDed with A.) Is the statement $A(A + B) = A$ correct?

A	B	$A + B$	$A(A + B)$
0	0		
0	1		
1	0		
1	1		

14A Yes

A	B	$A + B$	$A(A + B)$
0	0	0	0
0	1	1	0
1	0	1	1
1	1	1	1

15 **Theorem**

$$A(A + B) = A$$

16 Multiplying two expressions in Boolean algebra is very similar to multiplying two expressions in conventional algebra. The Boolean theorem governing multiplication is:

Theorem

$$A\,(B + C) = AB + AC$$

It can be proved by using the truth table. Basically, this theorem states that the distributive law holds in Boolean algebra, as it does in conventional algebra. After completing the multiplication, the letters are rearranged in alphabetical order. (This rearrangement is legal because $AB = BA$.)

Example 1

$$(A + \overline{B} + C)\,(E + F) = AE + AF + \overline{B}E + \overline{B}F + CE + CF$$

Example 2

If a variable is multiplied by itself, the result is not the square of the variable but the variable itself, because $A \cdot A = A$.

$$A\,(A + B + C) = A + AB + AC$$

17Q Multiply the following example out.

$(A + B)\,(C + \overline{D}) =$ ____________

17A $(A + B)\,(C + \overline{D}) = AC + A\overline{D} + BC + B\overline{D}$

18Q Multiply the expressions and remove parts which are 0.

$(AB + C)\,(\overline{B} + \overline{C}) =$ ____________

18A $AB\overline{C} + \overline{B}\,C$

Multiplying out gives

$$\underline{AB\overline{B}} + AB\overline{C} + \overline{B}C + \underline{C\overline{C}}$$

The underlined parts are 0 because in each a variable is multiplied by its inverse.

19Q Multiply these two expressions together and remove parts which are 0:

$$(\overline{A} + \overline{B} + \overline{C})\,(A + B + C) = \underline{\qquad\qquad}$$

19A $\underline{A\overline{A}} + \overline{A}B + \overline{A}C + A\overline{B} + \underline{B\overline{B}} + \overline{B}C + A\overline{C} + B\overline{C} + \underline{C\overline{C}} =$

$\overline{A}B + \overline{A}C + A\overline{B} + \overline{B}C + A\overline{C} + B\overline{C}$

Three terms are equal to zero and are therefore dropped out (underlined).

---

20 **Summary**

In Boolean algebra multiplying is the same as in conventional algebra with the exception that $A \cdot A = A$. Simplification is often possible owing to the theorem $A \cdot \overline{A} = 0$.

---

21 Before continuing with the use of the theorems in equations, let us define the notion of function. This will bring together several concepts and be very useful later.

Recall: A Boolean variable is a symbol which may take on two values (0 or 1). (Chap. 2, frame 1, page 18.)

Function: Strictly speaking, a function consists of three things:

1. A set of variables, called independent variables. (In this book the beginning of the alphabet is used.)

2. Another variable, called the dependent variable. (In this book the end of the alphabet is used.)

3. A rule which assigns a value to the dependent variable for every set of values of the independent variables.

Notation: The mathematical notation for a function with the independent variables A, B, and C, the dependent variable X, and the rule F is

$$X = F(A, B, C)$$

Example of a function: The equation $X = AB + C$ is a function in which A, B, and C are the independent variables, X is the dependent variable, and the rule is: "AND the values for A and B and OR the result with the value of C to get the value of X." For example, the rule states that

if	$A = 0$, $B = 0$, and $C = 0$, then $X = 0$
or if	$A = 0$, $B = 1$, and $C = 1$, then $X = 1$
or if	$A = 1$, $B = 1$, and $C = 0$, then $X = 1$

22Q Let us make use of the theorems proved earlier in this section. Assume that the theorem is broken into the two sides and each of the sides is set equal to a dependent variable. For example, the theorem $A(A + B) = A$ can be written as

$$X = A(A + B) \qquad \text{and} \qquad X = A$$

Note that the two values of X are the same. Implement the first form for X in AND and OR circuits. Start to implement the circuit with an AND. The output of the AND is X. Find the inputs to the AND and implement them.

22A

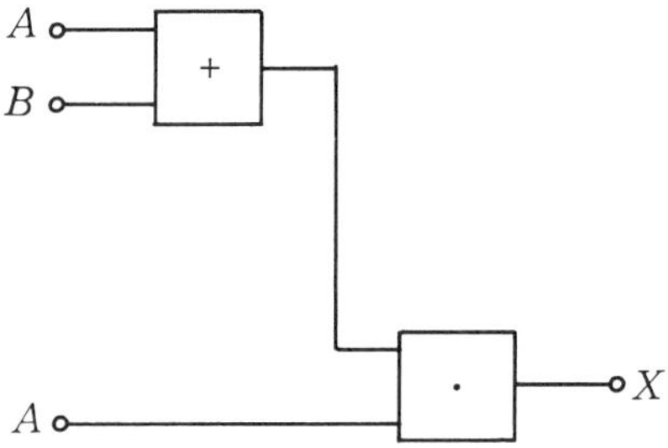

23 According to the theorem $A(A + B) = A$, X can also be implemented by a piece of wire from A to X.

$$A \text{ ———————— } X$$

This circuit (a short circuit) and the one in frame 22A are equivalent. If cost is a consideration it is needless to say that the second form of X is better. But before going into cost considerations let us exercise the design of equivalent circuits.

24Q

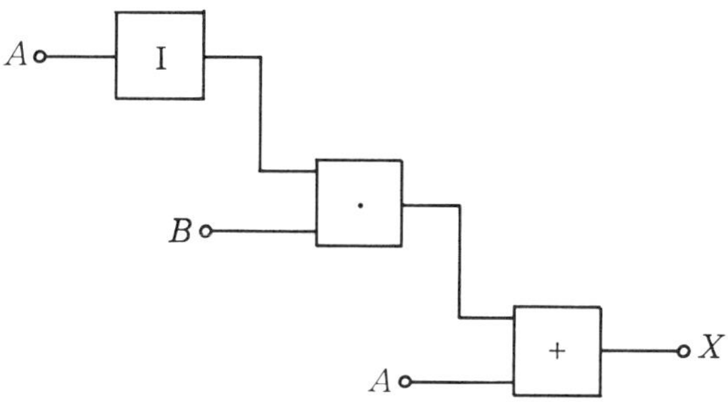

The equation for X = ____________.

24A $X = A + \bar{A}B$

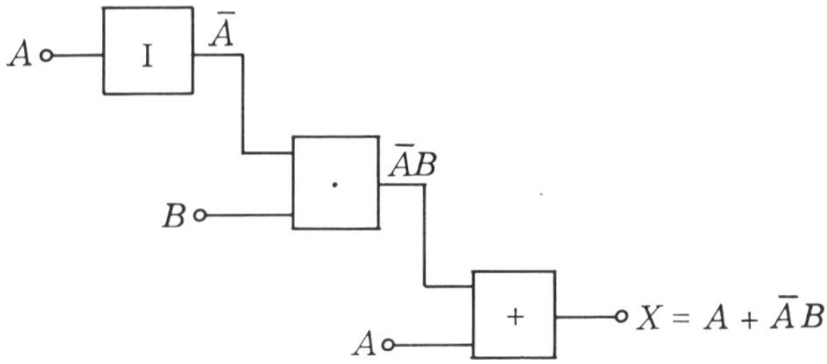

25Q Using the theorem $A + \bar{A}B = A + B$, an equivalent circuit can be developed. Make a sketch of it.

25A

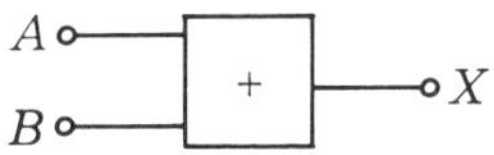

26 The comparison of the two circuits demonstrates the powerful use of a theorem.

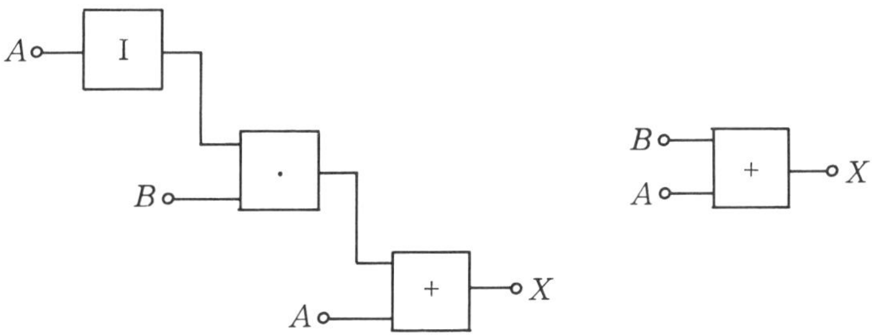

Original circuit	Equivalent circuit
1 INVERTER, 1 AND, and 1 OR	1 OR

27Q It is becoming very evident that there are equivalent circuits (see frame 26). For some applications it is not necessary to make a choice as to which one to use, but in the design of computers or control circuits, a choice has to be made. The basis for this choice, as in most engineering problems, is *cost*. It was pointed out in Chap. 1 that reducing cost is one of the objectives of this book.

It is not possible to put a dollar-and-cents value on each of the circuits, but as you will recall, both AND and OR circuits were made with diodes. In fact, each of the inputs for either one of the circuits was a diode. Thus the "count of diodes" of one circuit, as compared with another circuit, can be used as a measure of cost. INVERTERS, you recall, are made of transistors, which are usually more expensive than diodes. It is again not possible to lay down a simple relationship, such as two diodes are equivalent to one transistor. Therefore the diodes and transistors are counted separately.

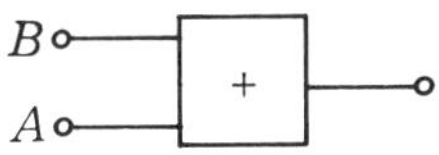

This circuit uses ______ transistor(s) and ______ diode(s).

27A 0 transistors; 2 diodes

Each input into the OR requires one diode; thus there are two diodes. As there is no INVERTER, no transistor is needed.

28Q

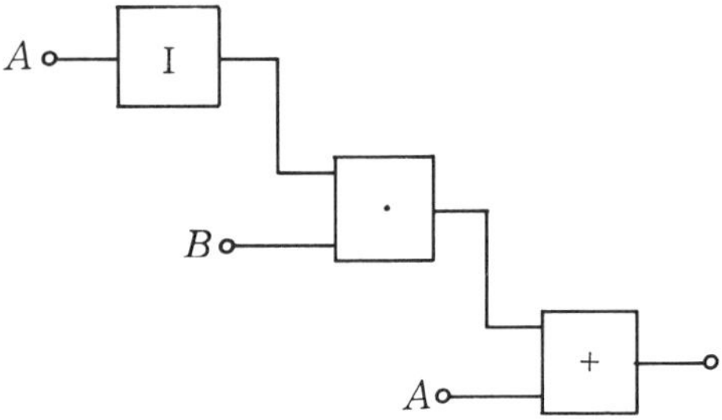

The cost of this circuit is _____ transistor(s) and _____ diode(s).

28A 1 transistor; 4 diodes

There is one INVERTER, and thus one transistor. Each input into an OR or an AND requires a diode; thus two diodes are needed for the OR and two for the AND, making a total of four diodes.

29 From frame 26 you recall that the two circuits were equivalent. The cost of each circuit is written under each diagram. From these figures we arrive at the same conclusion that the circuit on the right is a better choice for implementing the expression $\bar{A}B + A$ (or its equivalent expression).

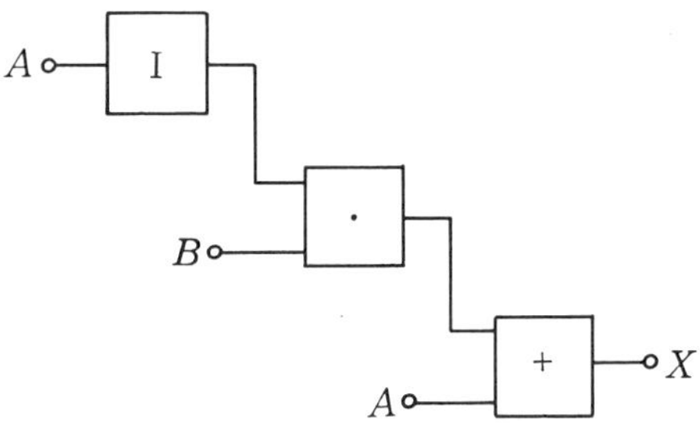

1 transistor, 4 diodes

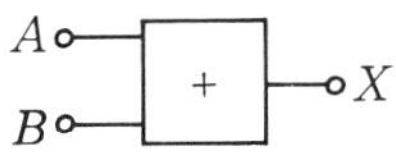

0 transistors, 2 diodes

30Q $X = AC + \overline{A}B + \overline{B}$. You are now in a position to use the theorems to improve circuits. Note that the expression for X is given in the form of an equation. X, the dependent variable, is the desired output of the circuit. A, B, and C, the independent variables, are the inputs to the circuit. Perform these steps:

1. Simplify the equation by using the theorems.

2. Implement the simplest form found in step 1.

Note: A Summary of Theorems is given at the end of the book. The theorem which applies here is $A\overline{B} + B = A + B$. This problem does not use the theorem in the form shown here or in the summary at the end of the book.

30A $X = \overline{A} + \overline{B} + C$

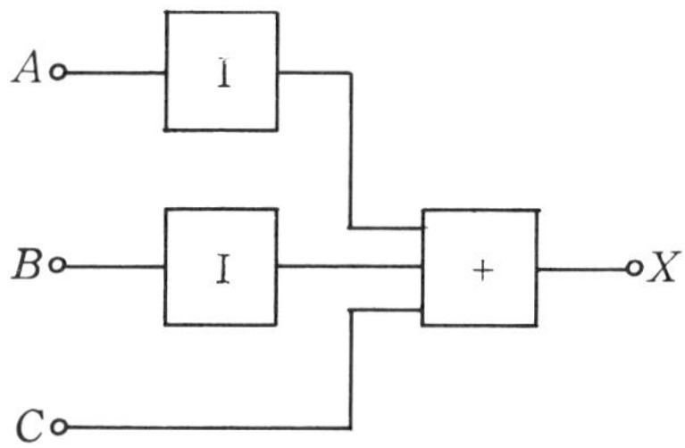

$$
\begin{aligned}
X &= AC + \overline{A}B + \overline{B} \\
&= AC + \overline{A} + \overline{B} && \text{(because } \overline{A}B + \overline{B} = \overline{A} + \overline{B}\text{)} \\
&= C + \overline{A} + \overline{B} && \text{(because } AC + \overline{A} = \overline{A} + C\text{)}
\end{aligned}
$$

31Q Implement X at the lowest possible cost.

$X = A\overline{B} + B + \text{AC}$

31A $X = \underbrace{A\overline{B} + B} + AC \qquad (A\overline{B} + B = A + B)$

$= A + B + AC \qquad (A + AC = A)$

$= A + B$

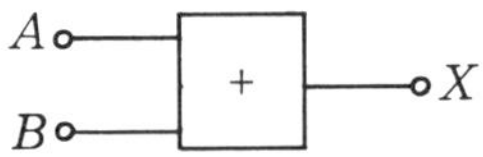

32 Thus far the variables in each theorem were assumed to be only simple variables. These same theorems can be applied to a large number of problems if parts of the expression under investigation are replaced by new variables which in turn are used in the already familiar theorems. This "replacement" of expressions by a single variable is an application of the substitution theorem. (Chap. 3, frame 40, page 49.)

Example

Reduce the expression

$$E(AB + CD) + \overline{E}$$

Replacing $AB + CD$ by Z permits the expression to be rewritten as

$$EZ + \overline{E}$$

which is equal to $Z + \overline{E}$. The original expression reduces to

$$(AB + CD) + \overline{E} \qquad \text{or} \qquad AB + CD + \overline{E}$$

The substitution permits us to apply simple theorems to relatively complex expressions.

33Q An extended notation for INVERSION will be introduced.

$$\overline{AB + C}$$

A bar over an entire expression (or part of an expression) means that the value of this expression (or part of the expression) will be the inverse of the original expression.

Example

If $W = A + CD$, it follows that $\overline{W} = \overline{A + CD}$.

Simplify the expression for X using the indicated substitutions.

$$X = (AB + C)(\overline{B}\,\overline{C} + D\overline{E}) + (\overline{AB + C})$$

Substitutions: $Y = AB + C$; $Z = \overline{B}\,\overline{C} + D\overline{E}$

33A

Given: $X = \underbrace{(AB + C)}(\underbrace{\overline{B}\,\overline{C} + D\overline{E}}) + \underbrace{(\overline{AB + C})}$

Substituting: $X = Y \cdot Z + \overline{Y}$

Simplifying: $X = Z + \overline{Y}$

Resubstituting: $X = \overline{B}\,\overline{C} + \mathrm{D}\overline{\mathrm{E}} + (\overline{AB + C})$

34Q Simplify the expression for U:

$$U = (A\overline{C}D + A\overline{B}\,\overline{C})(\overline{A\overline{C}D + A\overline{B}\,\overline{C}}) + \overline{A}\,\overline{B}$$

34A

Given: $U = (A\overline{C}D + A\overline{B}\,\overline{C})(\overline{A\overline{C}D + A\overline{B}\,\overline{C}}) + \overline{A}\,\overline{B}$

Assumed substitution: $X = A\overline{C}D + A\overline{B}\,\overline{C}$

Substitute: $U = \underbrace{X \cdot \overline{X}} + \overline{A}\,\overline{B}$

Simplify: $U = 0 + \overline{A}\,\overline{B} = \overline{A}\,\overline{B}$

Answer: $U = \overline{A}\,\overline{B}$

35Q Simplify the expression for Z:

$$Z = AB\overline{C} + D + E(AB\overline{C} + D)$$

35A

Given:	$Z = AB\overline{C} + D + E(AB\overline{C} + D)$
Assume:	$X = AB\overline{C} + D$
Substitute:	$Z = X + EX$
Simplify:	$Z = X$
Answer:	$Z = AB\overline{C} + D$

36 The substitution theorem is very useful in reducing expressions algebraically. It can also be used in simplifying circuits directly. For example (work left to right),

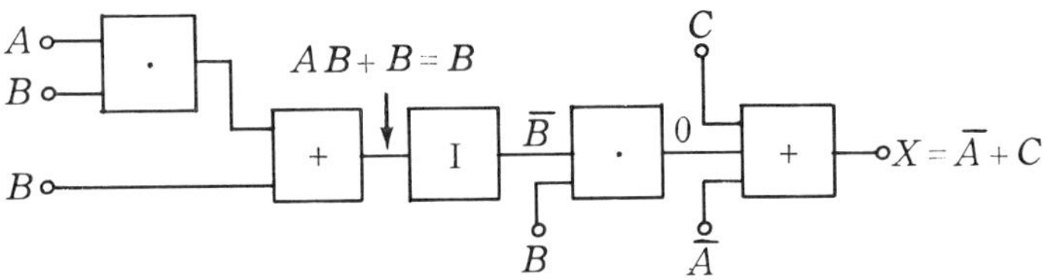

Additional problems of this type are included at the end of the chapter which follows.

PROBLEMS

1. Using the truth table, prove that $(A + B)(A + \overline{B}) = A$.

2. The theorem $A + \overline{A}B = A + B$ can be proved by examining the equality for all values of A and B. Reasoning the theorem out is equivalent to using the truth table. Assume $A = 1$; if $A = 1$,

then $\bar{A} = 0$, and the left side is $A + \bar{A}B = 1 + 0 \cdot B = 1$. The other side is $1 + B = 1$. Thus if $A = 1$, the equality holds. Assume $A = 0$; if $A = 0$ then $\bar{A} = 1$ and the left side is $A + \bar{A}B = 0 + 1 \cdot B$. The other side is $0 + B = B$. Thus if $A = 0$ the equality holds. The equality holds for $A = 0$ and $A = 1$; after trying the same with B the equality has been proved. Prove the equality $A + AB = A$ in the same manner that $A + \bar{A}B = A + B$ has been proved.

3. Evaluate the expressions:

(a) $1 \cdot \bar{0} \cdot \bar{1} + A\bar{B} + \bar{0} \cdot \bar{0} =$ ____________ .
(b) $(0 + \bar{1} + \bar{0})(\bar{A} + B)(\bar{1} + \bar{1}) =$ ____________ .

(Where possible, simplify the expressions in the parentheses first.)

4. Multiply the following expression, and simplify wherever possible:

$$(A + \bar{B} + \bar{C})(A + \bar{B} \cdot C) = \text{\_\_\_\_\_\_\_\_\_\_\_\_} .$$

5. Find the output of this circuit, and design a simpler circuit having the same output. Find the circuit output by first finding the output of each block, going from left to right, and simplifying as you go along.

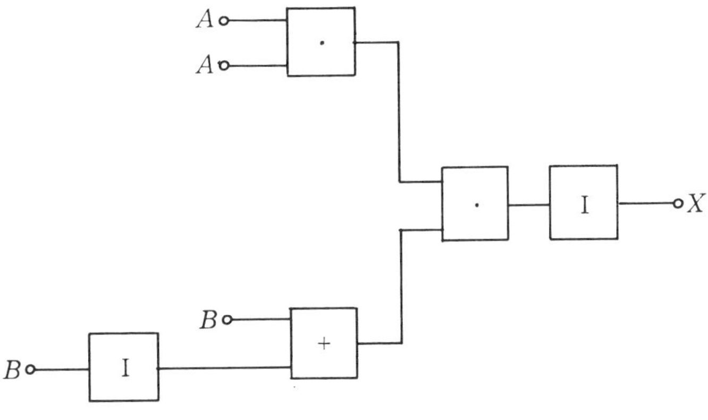

6. Simplify this circuit (see Prob. 5 for details).

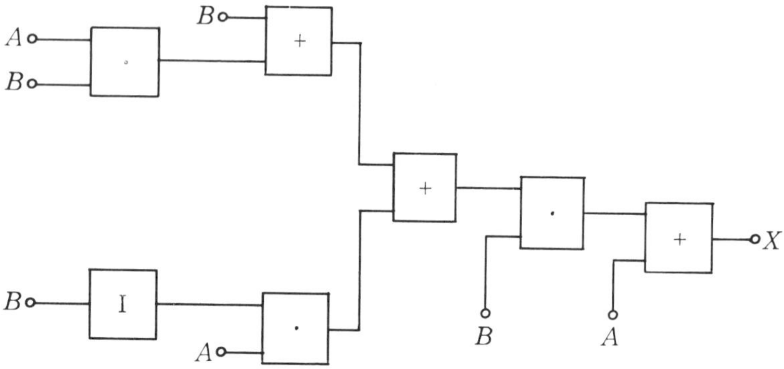

7. Simplify this circuit.

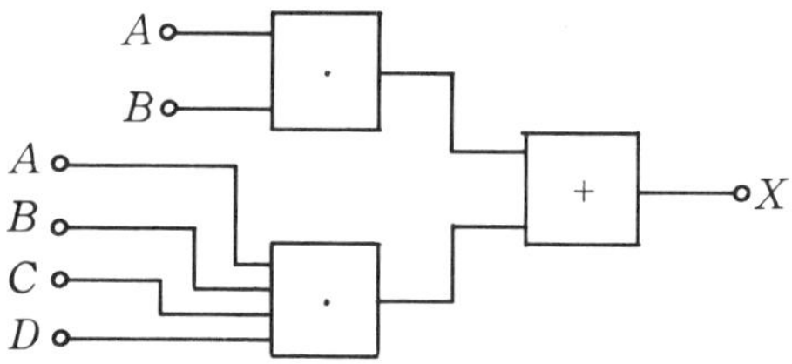

8. A bar can extend over several variables. You do not know how to break down such a bar, so do *not* break it down but handle it as a unity. For example $(\overline{AB + C})$ merely stands for the inverse of $(AB + C)$.

(*a*) Simplify the expression $(A\bar{B} + B)(\overline{A + B})$.

(*b*) Simplify the expression $B(\overline{C + D}) + AB(\overline{C + D})$.

(*c*) Simplify the expression $A + \bar{C} + B(\overline{A + \bar{C}}) + (A + \bar{C})(A + B + \bar{C})$.

Chapter 5

INVERSION

Chapter 2 described the inversion of a single constant or variable. In this chapter we go on to learn how to invert complete expressions. At first, the truth table will be used for proving that one expression is the inverse of another expression. De Morgan's theorems for inverting expressions are then introduced, and the method of using these theorems is given. Many examples are included, leading up to expressions having multiple inversions and requiring repeated use of the theorems. It is then shown how the theorems may be used to redesign and simplify circuits.

---

1 The ability to invert an expression has a number of important practical applications because there are always two ways to design a circuit for an expression. One way to design a circuit is by using the expression directly; the other way is by using the inverse of the expression and then adding an inverter block at the output of the circuit. It is often found that the indirect method results in a cheaper and therefore better circuit. A similar application occurs when a circuit must perform the inverse of a given expression. In this case, it may be better to invert the expression before designing the circuit than to use the original expression and add an inverter at the output of the circuit.

2Q *Inversion* has been defined as follows:

$\overline{0} = 1 \qquad \overline{1} = 0$

For each combination of the inputs (A and B) our "expression" takes on one value (0 or 1). According to the definition above the "inverse of that expression" will take on the opposite value. Complete the table.

A	B	Expression	Inverse of expression
0	0	1	
0	1	1	
1	0	0	
1	1	1	

2A

A	B	Expression	Inverse of expression
0	0	1	0
0	1	1	0
1	0	0	1
1	1	1	0

The "expression" is 1 in three rows. In these three rows the "inverse of the expression" has to be 0. In the remaining row the "expression" is 0; thus its inverse is 1.

3Q Inversion is indicated by a bar. The length of the bar determines what part of the expression is inverted. If, for example, the bar extends over the *entire* expression, the *entire* expression is to be inverted. Complete the table.

A	B	AB	$\overline{AB}$
0	0	0	
0	1	0	
1	0	0	
1	1	1	

3A

A	B	AB	$\overline{AB}$
0	0	0	1
0	1	0	1
1	0	0	1
1	1	1	0

If the expression is 0, its inverse is 1. Thus in the first three rows the entry is a 1. If the expression is 1, its inverse is 0, and so the entry for the last row is 0.

4Q Try this one by completing the table:

A	B	$A\overline{B} + \overline{A}B$	$\overline{A\overline{B} + \overline{A}B}$
0	0	0	
0	1	1	
1	0	1	
1	1	0	

4A

A	B	$A\overline{B} + \overline{A}B$	$\overline{A\overline{B} + \overline{A}B}$
0	0	0	1
0	1	1	0
1	0	1	0
1	1	0	1

5Q Let us return to the single-variable theorems. Complete the table.

A	$\overline{A}$
0	
1	

5A

A	$\overline{A}$
0	1
1	0

6Q Now that $\overline{A}$ has been found, find the entries for the inverse of $\overline{A}$ which is $\overline{\overline{A}}$.

$\overline{A}$	$\overline{\overline{A}}$
1	
0	

6A

A	$\overline{A}$	$\overline{\overline{A}}$
0	1	0
1	0	1

7 From the table in frame 6A you can see that the column A is identical to that for $\overline{\overline{A}}$. In equation form:

Theorem

$$A = \overline{\overline{A}}$$

The theorem states that, if a variable (expression) which is already inverted is inverted a second time, the initial variable (expression) is obtained; or double inversion does not change an expression and therefore can be dropped.

This frame concludes the general discussion of inversion. The next topic will be to develop the inversion theorem.

8Q Let us next consider the inversion of an expression having two variables. The expression $A + B$ and its inverse, $\overline{A + B}$, are plotted on the truth table. Is $\overline{A + B}$ equal to $\overline{A}\overline{B}$? (Use the table.)

A	B	$A + B$	$\overline{A + B}$
0	0	0	1
0	1	1	0
1	0	1	0
1	1	1	0

8A Yes

A	B	$\overline{A + B}$	$\overline{A}\overline{B}$
0	0	1	1
0	1	0	0
1	0	0	0
1	1	0	0

9Q Using the three-variable truth table prove that $\overline{A + B + C}$ is equal to $\bar{A}\bar{B}\bar{C}$.

A	B	C	$A + B + C$	$\overline{A + B + C}$	$\bar{A}\bar{B}\bar{C}$
0	0	0			
0	0	1			
0	1	0			
0	1	1			
1	0	0			
1	0	1			
1	1	0			
1	1	1			

9A

A	B	C	$\overline{A + B + C}$	$\bar{A}\bar{B}\bar{C}$
0	0	0	1	1
0	0	1	0	0
0	1	0	0	0
0	1	1	0	0
1	0	0	0	0
1	0	1	0	0
1	1	0	0	0
1	1	1	0	0

10 Generalizing the findings from the last two frames gives the general form of the theorem.

Theorem

$$\overline{A + B + C + \cdots} = \bar{A}\bar{B}\bar{C} \cdots$$

This theorem is called *De Morgan's theorem;* it defines inversion of expressions. In frame 12 this same theorem will be written in a different form.

11Q Apply De Morgan's theorem ($\overline{A + B + C} = \overline{A}\overline{B}\overline{C}$) to the expression

$$\overline{\overline{D} + E + F} = \_\_\_\_\_\_\_\_$$

11A $D\overline{E}\overline{F}$

De Morgan's theorem produces $\overline{\overline{D}}\overline{E}\overline{F}$, but since $\overline{\overline{D}} = D$, $\overline{\overline{D}}\overline{E}\overline{F} = D\overline{E}\overline{F}$.

12 The other form of De Morgan's theorem is:

Theorem

$$\overline{ABC \cdots} = \overline{A} + \overline{B} + \overline{C} + \cdots$$

This form of the theorem can also be proved by means of a truth table (see Prob. 5 at the end of the chapter).

13 **Summary**

The two forms of De Morgan's theorem are:

$$\overline{A + B + C + \cdots} = \overline{A}\overline{B}\overline{C} \cdots$$

$$\overline{ABC \cdots} = \overline{A} + \overline{B} + \overline{C} + \cdots$$

These equations show that the inverse of an expression of individual variables ANDed or ORed together is obtained by

1. Inverting all variables. (But if the variable was originally inverted, it is changed back to true form.)

2. Changing the OR to AND or the AND to OR.

14 Suppose a more complex expression is to be inverted, such as $(AB + C)$; neither form of De Morgan's theorem applies directly, but if AB is replaced by X, one form of the theorem applies. If $AB = X$,

$$\overline{AB + C} = \overline{X + C}$$
$$= \overline{X}\overline{C}$$

Resubstituting,

$$\overline{AB + C} = \overline{AB}\overline{C}$$

Carrying out the inversion of $\overline{AB}$ gives $\overline{A} + \overline{B}$.

Here is where most *errors* are made. All of $\overline{X}$ is multiplied by $\overline{C}$. Therefore, if $\overline{AB}$ is replaced by $\overline{A} + \overline{B}$, parentheses have to be placed around it. The final form of the expression is:

$$\overline{AB + C} = (\overline{A} + \overline{B})\overline{C}$$

A shorthand notation is introduced: Instead of replacing a part of an expression by a new dependent variable, braces are used to indicate what parts are taken as a unit. In this example, instead of replacing AB by X, the notation $\underbrace{AB}$ is used, which means AB is a unit which is not disturbed during this application of De Morgan's theorem. Reviewing the problem in steps.

$\overline{\underbrace{AB} + \underbrace{C}}$	apply theorem $\overline{A + B} = \overline{A}\overline{B}$
$\overline{\underbrace{A}\underbrace{B}}\overline{C}$	apply theorem $\overline{AB} = \overline{A} + \overline{B}$
$(\overline{A} + \overline{B})\overline{C}$	remember parentheses

15Q To invert the expression that follows, it can be broken into the parts as indicated by the braces. Apply De Morgan's theorems. You may want to use the Summary of Theorems at the end of the book as reference.

$$\overline{\underbrace{\overline{A}\overline{C}} + \underbrace{B} + \underbrace{AD}}$$

15A $\overline{\underbrace{\bar{A}\bar{C}} + \underbrace{B} + \underbrace{AD}}$

$\overline{\bar{A}\bar{C}} \cdot \bar{B} \cdot \overline{AD}$

The theorem which applies is $\overline{A + B + C} = \bar{A}\bar{B}\bar{C}$.

16Q Continue this problem by inverting the part indicated by the braces.

$$\overline{\underbrace{\bar{A}}\,\underbrace{\bar{C}}} \cdot \bar{B} \cdot \overline{AD}$$

16A $(A + C)\bar{B} \cdot \overline{AD}$

Did you remember the parentheses? The entirety of $\overline{\bar{A} \cdot \bar{C}}$ is multiplied by $\bar{B}$ and $\overline{AD}$. To retain this relationship the parentheses have to be added.

17Q Repeat the same step as above on the last part of the expression

$$(A + C)\bar{B} \cdot \underbrace{\overline{AD}}$$

17A $(A + C)\bar{B}(\bar{A} + \bar{D})$

Note the parentheses and the inversion of each variable.

---

18 You may note that parentheses should always be added whenever De Morgan's theorem is used in the form

$$\overline{ABC\cdots} = \bar{A} + \bar{B} + \bar{C} \cdots$$

Be liberal with parentheses; extra ones never make for errors, but missing ones account for a large part of errors in logic design.

19Q Invert the expression

$$(A+B+C)D$$

19A

$$\overline{(A+B+C)\ D}$$

$$\overline{(A+B+C)}+\bar{D} \qquad \text{Parentheses can be omitted}$$

$$\bar{A}\ \ \bar{B}\ \ \bar{C}+\bar{D} \qquad \text{[answer]}$$

20Q Invert the entire expression

$$\bar{A}B+\bar{C}\bar{D}+A\bar{C}+ACD$$

20A

$$\overline{\bar{A}B+\bar{C}\bar{D}+A\bar{C}+ACD}$$

$$\overline{\bar{A}B}\cdot\overline{\bar{C}\bar{D}}\cdot\overline{A\bar{C}}\cdot\overline{ACD}$$

$$(A+\bar{B})(C+D)(\bar{A}+C)(\bar{A}+\bar{C}+\bar{D}) \qquad \text{[answer]}$$

21Q This next expression will require a few more steps. Carry out the inversion

$$\overline{(A+B)\bar{C}+(\bar{A}+\bar{D})B}$$

21A

$$\overline{(A+B)\,\bar{C} + (\bar{A}+\bar{D})\,B}$$

$$\overline{(A+B)\,\bar{C}} \cdot \overline{(\bar{A}+\bar{D})\,B}$$

$$[\overline{(A+B)} + C] \cdot [\overline{(\bar{A}+\bar{D})} + \bar{B}]$$

$$[\bar{A}\;\bar{B} + C] \cdot [A\;D + \bar{B}] \qquad \text{[answer]}$$

22Q Carry out this inversion

$$\overline{[(A+B)C+D]E+F}$$

22A

$$\overline{[(A+B)\;C+D]\;E+F}$$

$$\overline{[(A+B)\;C+D]\;E}\;\bar{F}$$

$$\{\overline{[(A+B)\;C+D]}+\bar{E}\}\;\bar{F}$$

$$\{[\overline{(A+B)\;C}\cdot\bar{D}]+\bar{E}\}\;\bar{F}$$

$$\{[((\overline{A+B})+\bar{C})\;\bar{D}]+\bar{E}\}\;\bar{F}$$

$$\{[((\bar{A}\;\bar{B})+\bar{C})\;\bar{D}]+\bar{E}\}\;\bar{F} \qquad \text{parentheses eliminated}$$

$$\{[(\bar{A}\;\bar{B}+\bar{C})\;\bar{D}]+\bar{E}\}\;\bar{F} \qquad \text{[answer]}$$

23 Summary

Inverting an expression is done by repetitively applying one of the two forms of De Morgan's theorem. These forms are:

(*a*) $\overline{A+B+C+\cdots} = \bar{A}\bar{B}\bar{C}\cdots$

(*b*) $\overline{ABC\cdots} = \bar{A}+\bar{B}+\bar{C}+\cdots$

Form (*a*) often makes an existing set of parentheses unnecessary; form (*b*) usually requires *new* sets of parentheses at the ends of each bar which indicates inversion.

For example,

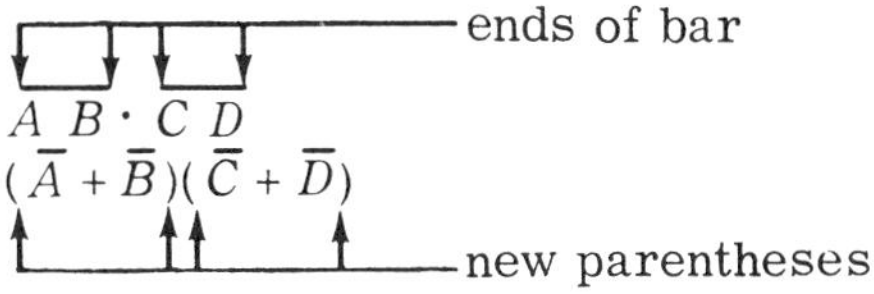

It is correct to omit the parentheses in this expression:

$$(\overline{A+B}) + \overline{C} = \overline{A+B} + \overline{C}$$

The reason is that the bar indicates that the expression $A + B$ is to be inverted and the result ORed with $\overline{C}$. The parentheses around $(A + B)$ merely emphasizes this point. Although it is correct to leave them out in a case like this, they will usually be used to increase clarity.

24Q In the next two problems pay particular attention to the correct use of parentheses. When in doubt, *use* them. Invert

$$(\overline{A}\,\overline{B} + C)\overline{E} + \overline{F}(\overline{A} + B + \overline{C}) + \overline{C}\,\overline{D} + EF$$

24A

$$\overline{(\overline{A}\,\overline{B}+C)\,\overline{E} + \overline{F}\,(\overline{A}+B+\overline{C}) + \overline{C}\,\overline{D} + E\,F}$$

$$\overline{(\overline{A}\,\overline{B}+C)\,\overline{E}} \cdot \overline{\overline{F}\,(\overline{A}+B+\overline{C})} \cdot \overline{\overline{C}\,\overline{D}} \cdot \overline{E\,F}$$

$$[\overline{(\overline{A}\,\overline{B}+C)}+E]\,[F+\overline{(\overline{A}+B+\overline{C})}]\,(C+D)\cdot(\overline{E}+\overline{F})$$ add parentheses

$$[(\overline{\overline{A}\,\overline{B}}\,\overline{C})+E]\,[F+A\,\overline{B}\,C]\,(C+D)\,(\overline{E}+\overline{F})$$ parentheses eliminated

$$[(A+B)\,\overline{C}+E]\,[F+A\,\overline{B}\,C]\,(C+D)\,(\overline{E}+\overline{F})$$ [answer]

25Q Carry out the indicated inversion. Note that only parts of the expression are to be inverted.

$$(\overline{AB} + C)\overline{D} + [\overline{D + \overline{A}\,\overline{B}C(E + \overline{D})}]$$

25A

$$(\overline{A \quad B} + C)\overline{D} + [\overline{D + \overline{A} \quad \overline{B} \quad C \quad (E + \overline{\overline{D}})}]$$

$$((\overline{\overline{A}} + \overline{\overline{B}}) + C)\overline{D} + [\overline{D} \cdot \overline{\overline{A} \quad \overline{B} \quad C \quad (E + \overline{\overline{D}})}]$$

$$(\overline{A} + \overline{B} + C)\overline{D} + [\overline{D} \cdot (A + B + \overline{C} + (\overline{\overline{E} + \overline{\overline{D}}}))]$$

$$(\overline{A} + \overline{B} + C)\overline{D} + \overline{D} \cdot (A + B + \overline{C} + \overline{\overline{E}} \quad D)$$ [answer]

26Q In some expressions, more than one inversion is indicated for part of an expression. In such a case any one of the inversions can be attacked first. If two bars are over the same part of an expression and do not extend beyond it they can be dropped because $\overline{\overline{A}} = A$. For example,

$$\overline{\overline{AB}}\overline{C} + D = AB\overline{C} + D$$

but

$$\overline{\overline{AB}C} + D \neq AB\overline{C} + D$$

To demonstrate this point, the expression below will be "untangled" by removing the long bar first. First remove the long bar, then the short one.

$$\overline{\overline{AB + C} + \overline{A}\overline{B} + D}$$

26A

$$\overline{\overline{AB + C} + \overline{A}\overline{B} + D}$$

$$(\overline{\overline{AB + C}}) \quad (\overline{\overline{A}\overline{B}}) \quad (\overline{D})$$

$$(AB + C) \quad (A + B) \quad (\overline{D})$$ [answer]

27Q Using the same expression as in frame 26, first remove the short bar, then the long one.

$$\overline{\overline{AB + C} + \overline{A}\overline{B} + D}$$

27A

$$\overline{\overline{A\quad B + C} + \bar{A}\quad\bar{B} + D}$$

$$\overline{\overline{A\;.\;B}\quad\bar{C} + \bar{A}\quad\bar{B} + D}$$

$$\overline{(\bar{A} + \bar{B})\quad\bar{C} + \bar{A}\quad\bar{B} + D}$$

$$[\overline{(\bar{A} + \bar{B})\quad\bar{C}}\;]\quad[\overline{\bar{A}\quad\bar{B}}\,]\quad[\bar{D}]$$

. . . .

$$[A\quad B + C]\quad[A + B]\quad[\bar{D}]$$

28 In the last two frames the same problem was solved two different ways. The two answers are the same. In general, if a problem has more than one inversion, the sequence in which these inversions are carried out is of no consequence. It should be pointed out that any intermediate result is equal to any other intermediate result and to the final answer.

This frame concludes the presentation of new material on inversion. If you are uncertain about inversion or have made more than three errors in last ten frames go back to frame 13 and repeat the problems.

29Q Carry out the various inversions in any order you choose.

$$\overline{\overline{AB}\,\overline{CD}}$$

29A $\overline{\overline{AB}\quad\overline{CD}}$ choose to work on long bar first

$\overline{\overline{AB}} + \overline{\overline{CD}}$

$AB + CD$ [answer]

30Q Write the expression for the output of each block and an equation for X. Rewrite X in such a way that no bar extends over more than one variable.

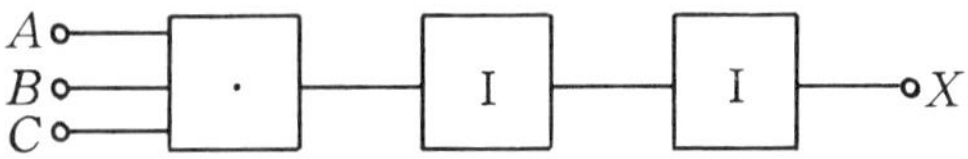

30A

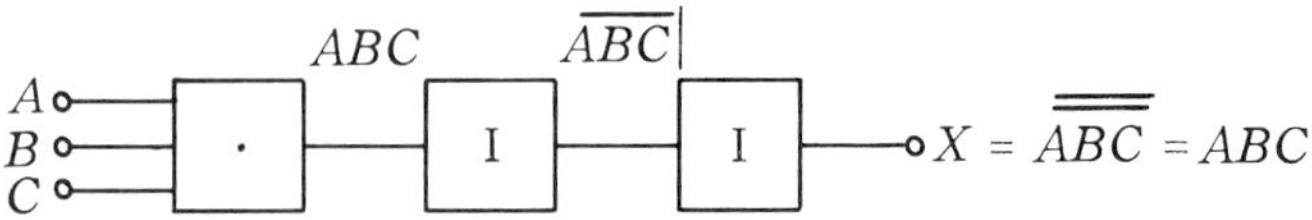

31Q Repeat the steps taken in frame 30 for this circuit.

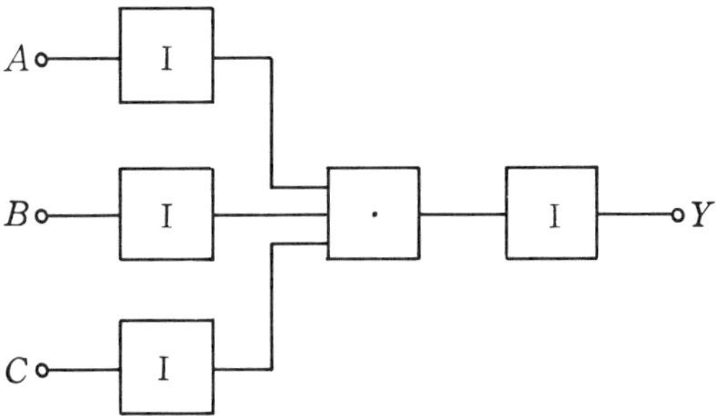

31A

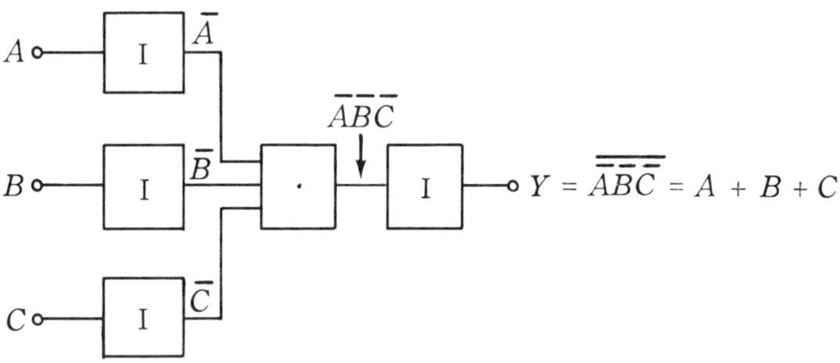

32Q Find Y in a form where no bar is longer than one variable.

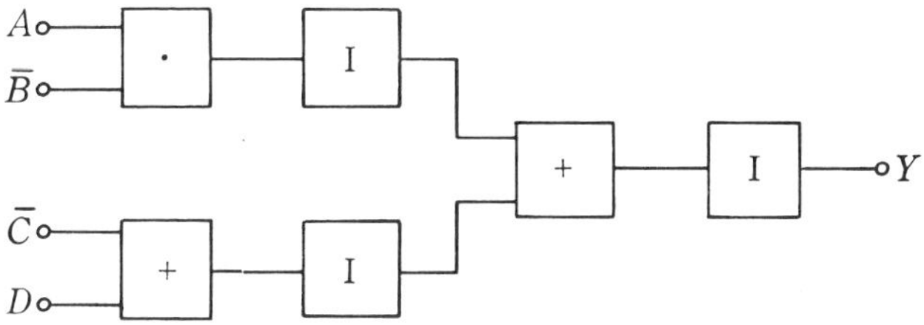

32A

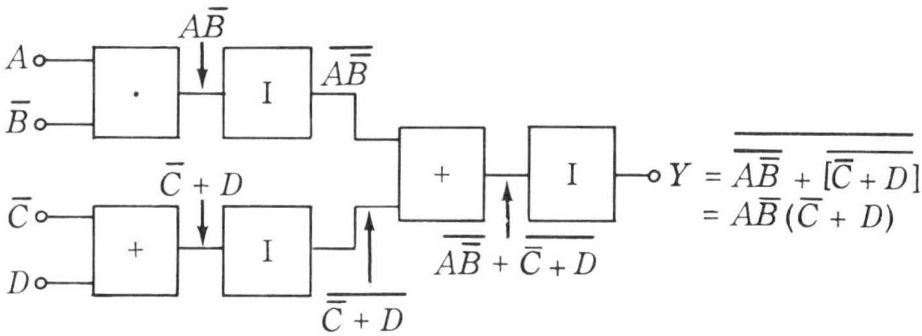

33Q Find X for this circuit in simplified form. (*Hint:* The simplified answer should contain four letters.)

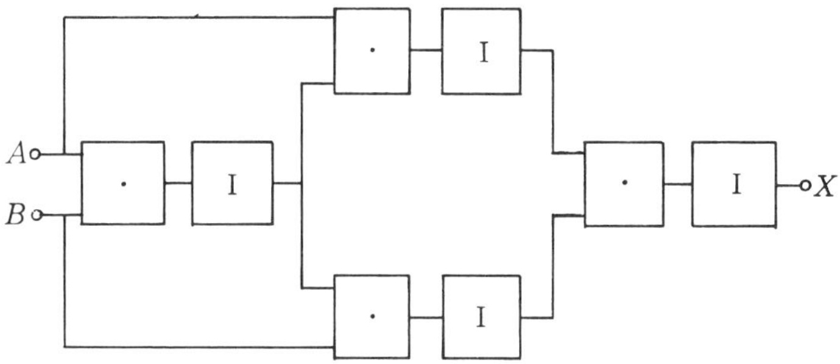

33A

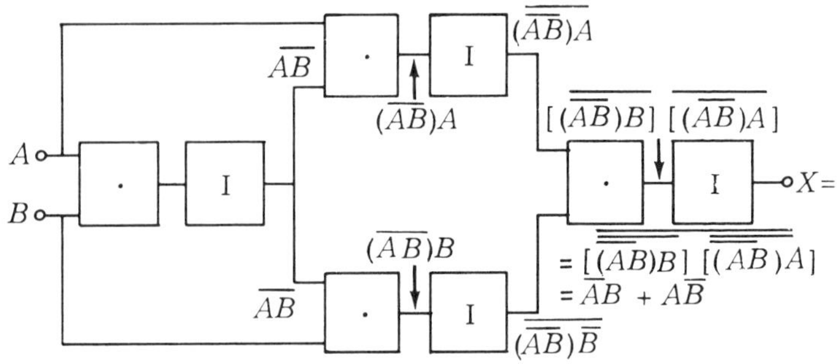

34Q Let us turn our attention away from inverting expressions and use inversion to prove additional theorems in Boolean algebra. By inverting both sides of a theorem which has already been proved, we have a new theorem which is necessarily correct. This is true because performing inversion on *both* sides changes *both* from 0 to 1 or 1 to 0. If the equality was true before inversion it must be true after the inversion. As an example, invert both sides of the theorem $A \cdot A = A$.

34A $\overline{A} + \overline{A} = \overline{A}$

35Q The original theorem ($A \cdot A = A$) was proved in Chap. 2. By inverting both sides, a new equation of the form $\overline{A} + \overline{A} = \overline{A}$ has been obtained. This constitutes a proof for the theorem $A + A = A$ because both are of the same form. One can be gotten from the other by substitution of variables.

The theorem below was proved in Chap. 4. By inverting both sides, prove another theorem included in the Summary of Theorems at the back of the book.

$$A + \overline{A}B = A + B$$

35A $\overline{A + \bar{A} B} = \overline{A+B}$

$\overline{A + (\bar{A} B)} = \overline{A+B}$

$\bar{A} \quad \overline{(\bar{A} B)} = \bar{A} \bar{B}$

$\bar{A} \quad (A+\bar{B}) = \bar{A} \bar{B}$

This is of the form

$$(A + \bar{B})B = AB$$

which is a theorem in the back of the book.

---

36 **Summary**

De Morgan's theorem

$$\overline{A + B + C + \cdots} = \bar{A}\bar{B}\bar{C} \cdots$$
$$\overline{ABC \cdots} = \bar{A} + \bar{B} + \bar{C} + \cdots$$

Any expression may be inverted by a repeated application of the two forms of this theorem.

PROBLEMS

1. Invert and carry out the inversion of the following expressions:

(*a*) $A + B + \bar{C}D$
(*b*) $\overline{AB} + C$
(*c*) $\overline{AB + \bar{C} + \bar{D}}$
(*d*) $A + B(\bar{C} + \bar{D})$
(*e*) $A(BC + \bar{D}) + A(\bar{B}\bar{C} + D)$

2. Implement the following expressions such that the last block in the circuit is an INVERTER. Do *not* use double inverters to satisfy the requirements.

(*a*) $(AB + C)$

(*b*) $(B + C)A + D$

3. Write the expression for the following circuit:

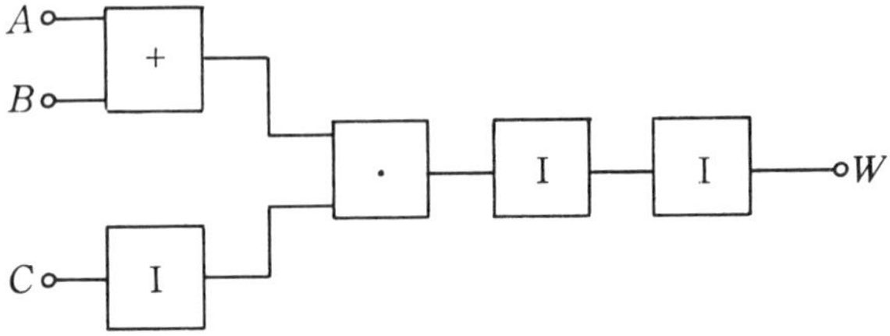

4. Write an equation for the following circuit without having (in the final answer) bars which extend over more than one variable.

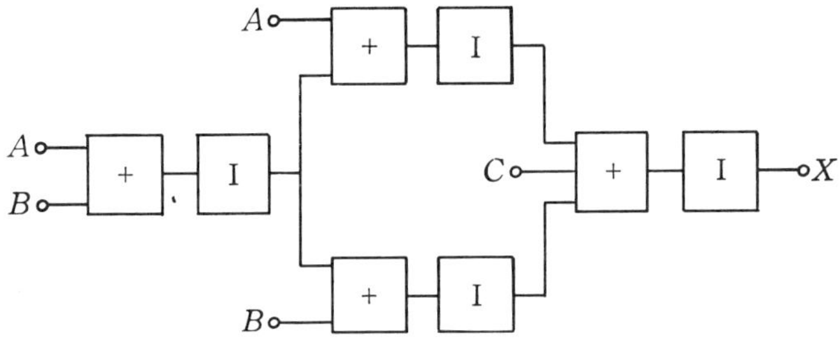

5. Prove $\overline{ABC} = \bar{A} + \bar{B} + \bar{C}$ on the truth table.

6. Find U and V and simplify.

(a)

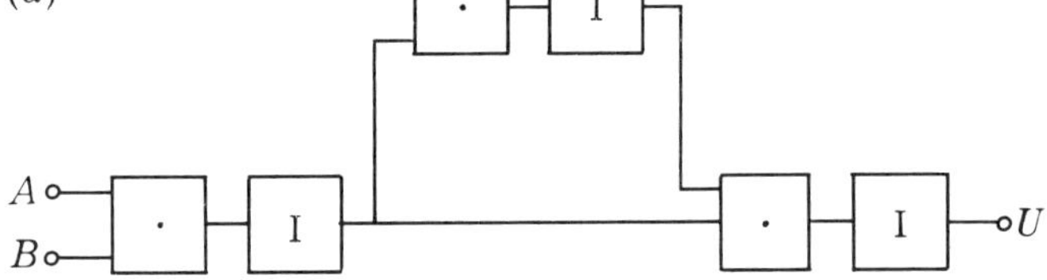

(b)

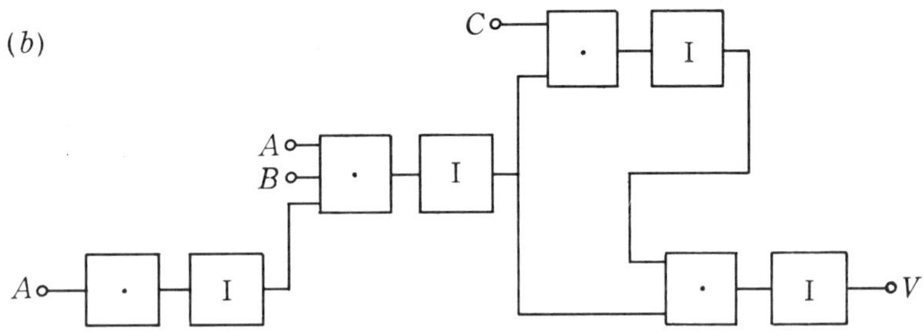

7. Equations are inverted by inverting both sides. Invert these equations:

   (*a*) $(A + B)(A + C) = A + BC$
   (*b*) $(A + B)(B + C)(\bar{A} + C) = (A + B)(\bar{A} + C)$

8. To prove that the equality is preserved during inversion, plot both sides of the equation in Prob. 7 on a truth table (before and after inversion).

   (*a*) Equation (*a*) from Prob. 7.
   (*b*) Equation (*b*) from Prob. 7.

Chapter 6

WORD PROBLEMS

The purpose of this chapter is to describe how an English sentence can be converted to a truth table or Boolean equation. At the beginning of the chapter, it is shown how a variable can be assigned to each phrase in the sentence. A truth table can then be written by evaluating the dependent variable for all conditions of the independent variables. It is also shown how an equation can be written according to the structure of the sentence. Using these techniques, circuits are designed to satisfy functions which are expressed in sentence form.

---

1 Normally, when a logic design problem is approached, it is not given in terms of a Boolean equation or a Boolean expression but in terms of one or more English sentences or statements. The first part of the problem is to break these sentences into parts (let us call them phrases) which are equivalent to a Boolean variable. The second part is to draw a truth table and plot the "truth" or "falsity" of the statements, thereby defining a Boolean function.

2Q First, the given statements have to be analyzed to find the phrases which are equivalent to Boolean variables.

Boolean variables can take on two values, 1 and 0. If some phrase is to be equivalent to a variable, it must have only two values. The values we have chosen are "true" and "false."

An example of a valid phrase is "Tom studies." This phrase is either true or false, depending upon Tom's activity at the time of the investigation.

An example of a sentence which does not qualify as a phrase is "eat vegetables," because no truth value is associated with this sentence.

Check those lines which correspond to a phrase which contains truth value.

1. He smokes a pipe.

2. Tomorrow is Sunday.

3. Smoke a pipe.

2A 1. He smokes a pipe. ✓

2. Tomorrow is Sunday. ✓

3. Smoke a pipe.

It is possible to agree or disagree with 1 and 2. As for 3, there is nothing to agree or disagree with.

3Q One sentence may contain several phrases. For simplicity's sake, the phrases will be marked with braces. For example,

If it rains tomorrow, I will study logic.

This sentence contains two phrases. The "if" is not part of either one, it merely shows the relationships.

Determine the phrases in the sentence below and mark them by braces.

The computer operates if cards are in the card reader.

3A The computer operates if cards are in the card reader.

It is either true or false that the computer operates, therefore the first three words are one phrase. The second half of the sentence can again be true or false; therefore, it, too, is a phrase.

4Q Each of the phrases can be set equal to a Boolean variable. In frame 3A above, variable A can be chosen to be equivalent to the phrase "The computer operates." Variable B can be chosen to be equivalent to the phrase "cards are in the card reader."

In the sentence below, three phrases appear. Find them, mark them with braces, and assign variables.

> John reads novels if it is an odd day of the month or it is Tuesday.

4A $\underbrace{\text{John reads novels}}_{A}$ if $\underbrace{\text{it is an odd day of the month}}_{B}$ or $\underbrace{\text{it is Tuesday}}_{C}$.

Note there are a few words which are not part of any phrase.

5Q Earlier the difference between dependent and independent variables was described. In review: The value of the dependent variable can be determined by assuming values for each of the independent variables. In this book, the letters at the end of the alphabet were chosen to represent dependent variables, letters at the beginning of the alphabet to represent independent variables.

In frame 4A, all variables were labeled as though they were independent, but one is a dependent one. Which is it?

5A Variable *A*

John's reading of novels depends on the truth or falsity of the other two phrases.

6Q A better assignment of variables for the sentence in frame 4A is

$$\underbrace{\text{John reads novels}}_{Z} \text{ if } \underbrace{\text{it is an odd day of the month}}_{B} \text{ or it is } \underbrace{\text{Tuesday.}}_{C}$$

Assign the proper variables to the phrases in the following sentence:

He watches TV if he has completed his studies or he is tired.

6A $$\underbrace{\text{He watches TV}}_{Z} \text{ if } \underbrace{\text{he has completed his studies}}_{A} \text{ or } \underbrace{\text{he is tired.}}_{B}$$

---

7 To fully specify a Boolean function, the relationship between the dependent and independent variables has to be established. First we have to be sure that we have all the information, then this information can be translated into Boolean notation. One such notation is the truth table.

To write the relationship between the dependent variable and the independent variables, all cases have to be defined. In the example in frame 6 not all were defined. Here is the problem: "He watches TV if he has completed his studies or if he is tired." This sentence states some of the cases why he watches TV, but nothing has been said that these are all the cases. To make the problem unique the *if* has to be replaced *if and only if*. Now it is absolutely certain when he does and when he does not watch TV. In the literature *if and only if* is often not written out, but written as *iff*. This notation will be adopted in the remainder of this book.

We are now ready to write a truth table for the example: He watches TV iff he has completed his studies or he is tired.

A	*B*	*Z*
0	0	
0	1	
1	0	
1	1	
Independent variables		Dependent variable

Rule: If a phrase is true, the variable associated with that phrase takes on the value 1.

To determine the value of *Z*, the truth or falsity of the associated phrase has to be examined for all values of the independent variables. In simple problems, this can be done by inspection. In complex problems, one row is examined at a time, and the value of the dependent variable determined.

8Q Example

$$\underbrace{\text{He watches TV}}_{Z} \text{ iff } \underbrace{\text{he has completed his studies}}_{A} \text{ or } \underbrace{\text{he is tired.}}_{B}$$

One condition which will cause him to watch TV is "he has completed his studies." Thus, if *A* is *true*, it follows that *Z* is *true*. $A = 1$ in the truth table corresponds to a true phrase; thus 1's are entered into the last two rows under the variable Z.

A	B	Z
0	0	
0	1	
1	0	1
1	1	1

Another condition causing him to watch TV is "he is tired." Repeat the above reasoning for this phrase and enter the result in the table.

8A

A	B	Z
0	0	
0	1	1
1	0	1
1	1	1

If B (he is tired) is 1, then Z (he watches TV) is 1.

9Q In frame 8A, the table contains three 1's under the Z.

What should be the entry in the first row, realizing that *all* conditions for which he watches TV are included in the initial sentence?

9A 0

Neither of the phrases A and B is true; thus the phrase "he watches TV" is false.

10Q Complete the following example by drawing a table and plotting the output Z.

> The TV set operates satisfactorily iff the set is turned on and the antenna is connected.

10A The TV set operates satisfactorily iff the set is turned on and the antenna is connected.

(Under-braces: "The TV set operates satisfactorily" = Z; "the set is turned on" = A; "the antenna is connected" = B.)

A	B	Z
0	0	0
0	1	0
1	0	0
1	1	1

Only when both A and B are true (1) can Z be true (1).

11 In the last few frames, the truth table was used to specify functions. This table constitutes a unique specification of the Boolean function. Using techniques covered in later chapters, the table can be read and the resulting equation simplified and implemented.

An alternate approach to specifying a Boolean function is to write an equation directly from the statement of the problem. For example,

John reads novels iff it is an odd day of the month or it is Tuesday.

(Under-braces: "John reads novels" = Z; "it is an odd day of the month" = A; "it is Tuesday" = B.)

$$Z = A + B$$

This equation says that if either $A = 1$ (it is an odd day of the month) or $B = 1$ (it is Tuesday), $Z = 1$ (John reads novels). Note that the equation defines that if either $A = 1$ *or* $B = 1$, or both, the variable $Z = 1$. This means that if it is both "an odd day of the month" and "a Tuesday," "John reads novels." This translation is the conventional one for the English conjunction "or."

12Q Translate the following sentence into an equation.

The TV set operates satisfactorily iff the set is turned on and the antenna is connected.

(Z = "The TV set operates satisfactorily"; A = "the set is turned on"; B = "the antenna is connected")

12A $Z = AB$

Z is true (1) if both A *and* B are true (1).

13Q Is the equation in frame 12A equivalent to the table plotted for this problem in frame 10A?

13A Yes

---

14 **Summary**

Problems are sometimes presented to the logic designer in the form of English sentences. To translate them into Boolean algebra, the two approaches which are summarized below are possible.

1. Phrases are formed and marked by braces.

2. Phrases corresponding to dependent and independent variables are determined and letters are assigned.

3. Either *a* or *b*:
   a. A truth table for the independent variables is drawn; the dependent variable is plotted on it.

b. An equation is written which connects the variables in the same manner in which the original phrases are connected by the English conjunctions.

15Q Let us next consider sentences which contain the word "not." Suppose

A stands for "It rains."
B stands for "It does not rain."

Assume $A = 1$, then the phrase "it rains" is true. At that time the phrase "it does *not* rain" must be false, the value of $B = 0$. Repeat this same reasoning for $A = 0$.

15A $A = 0$

"It rains" is false. "It does not rain" must be true. Value of B is 1.

16Q If $A = 0$, $B = 1$. If $A = 1$, $B = 0$. Write a Boolean relationship for the two variables.

16A $A = \overline{B}$ or $\overline{A} = B$

17Q In the above example, the B can be replaced by $\overline{A}$. If variable A stands for one phrase, $\overline{A}$ stands for the same phrase, but with a *not* added.

A = the set is turned on
B = the antenna is connected
Z = the TV set operates satisfactorily

Assuming the variables stand for the phrases without *not's*, as shown above, write an equation for this sentence:

The TV set does not operate satisfactorily iff the set is not turned on or the antenna is not connected.

17A $\overline{Z} = \overline{A} + \overline{B}$

Each phrase has a *not* which was not in the original phrase, thus each variable is inverted.

18Q Invert both sides of the equation in frame 17A.

18A $Z = AB$

$\overline{\overline{Z}} = \overline{\overline{A} + \overline{B}} \qquad Z = \overline{\overline{A}}\,\overline{\overline{B}} = AB$

---

19 Note that frames 18A and 12A are the same (frame 12A: $Z = AB$). Summarizing, adding a *not* to a phrase means inverting the variable standing for that phrase.

20Q You will probably not be familiar with the words used in this sentence because they are taken from computer terminology. Even though you are not familiar with the words, you should be able to write an equation from the structure of the sentence because in the translation, the phrases are changed to variables, the conjunctions to Boolean operations.

Write an equation for the sentence below:

> The sign of the accumulator is set positive iff a set positive instruction is given or a reset accumulator instruction is given.

20A The sign of the accumulator is set positive iff a set positive instruction is given or a reset accumulator instruction is given.

(Underbraces: "The sign of the accumulator is set positive" = Z; "a set positive instruction is given" = A; "a reset accumulator instruction is given" = B.)

$$Z = A + B$$

21Q Write an equation for this sentence.

Subtract iff an add instruction is given and the signs are different, or a subtract instruction is given and the signs are alike.

21A Subtract iff an add instruction is given and the signs are different or a subtract instruction is given and the signs are alike.

(Underbraces: "Subtract" = Z; "an add instruction is given" = A; "the signs are different" = B; "a subtract instruction is given" = C; "the signs are alike" = D.)

$$Z = AB + CD$$

22Q "Signs are alike" is the same as "signs are not different." Rewrite the sentence in frame 21A without using the variable D.

22A $D = \overline{B}$

$$Z = AB + C\overline{B}$$

23Q When it comes to implementing an equation in terms of logic blocks, we have been accustomed to labeling the inputs with the appropriate variables. Now that these variables are only an in-between step in the logic, we will not use the variables but the phrases which they stand for. We are actually defining input lines. For example, assume that the equation $Z = AB + C$ stands for the sentence:

$\underbrace{\text{Latch L is set}}_{Z}$ when (iff) $\underbrace{\text{the clock is at time 6}}_{A}$ and $\underbrace{\text{the computer is waiting for another instruction}}_{B}$, or $\underbrace{\text{when the computer is first turned on.}}_{C}$

Draw the logic blocks which would implement this equation and label all inputs and the output with the proper phrases, not the letters.

23A

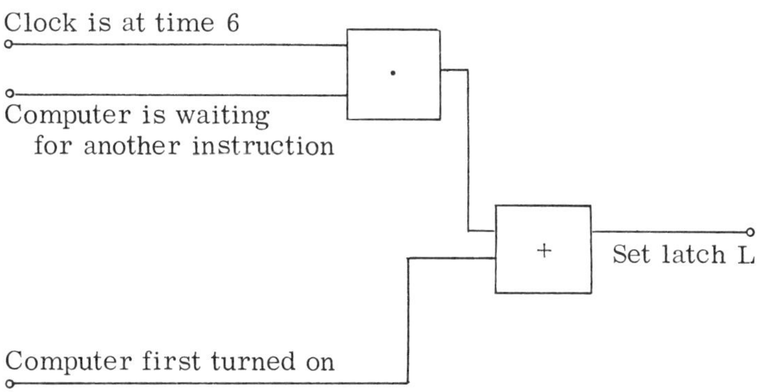

24Q Write an equation and draw a circuit for the following sentence:

The computer run light should be turned on when (iff) cards are read into the card reader and when the power is on, when the start button is pressed, or when no ready signal is coming from the console typewriter and the computer is ready.

24A The computer run light should be turned on when (iff)

Z

cards are read into the card reader and when the power

A *B*

is on when the start button is pressed or when no ready

C

signal is coming from the console typewriter and the

D

computer is ready.

E

$Z = AB + C + DE$

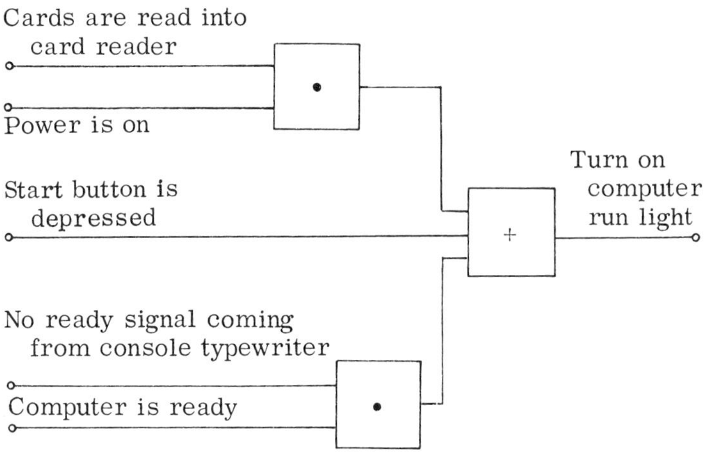

25Q We have assumed that the phrases correspond directly to available inputs to the circuit. This is not always the case, which means that the signal required by the circuit has to be generated from other related signals.

Example

Given: A wire labeled "accumulator zero," which means it is at the 1 level if the accumulator is 0.

Required: A wire labeled "accumulator not zero."

Find a circuit whose output is the required signal and whose input is the given signal. Draw a circuit and label inputs and outputs. *Hint:* First find a valid equation with the output on one side, a relationship of the available inputs on the other side.

25A

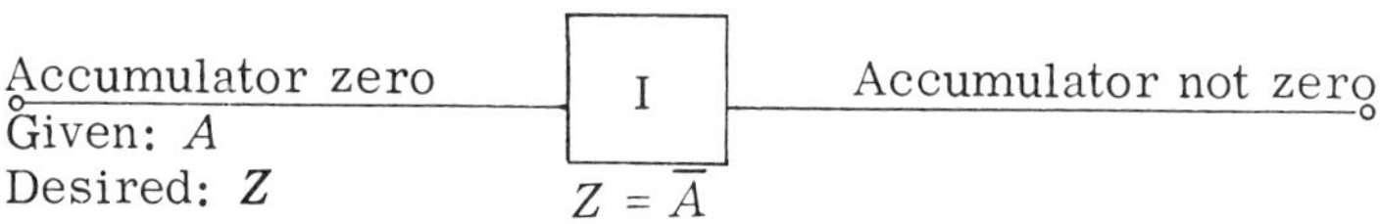

26Q A more complex example is given below. The available inputs and the desired output are listed. Check the equation for correctness, implement it, and label all inputs and the output of the circuit. Words in parentheses are added to make complete phrases, but are omitted in conventional English.

Given:

1. A wire labeled "accumulator positive or (accumulator) zero"
2. A wire labeled "accumulator zero"
3. A wire labeled "accumulator zero and (accumulator) positive"

Desired: A wire labeled "accumulator positive"

Assume: Variable A: accumulator positive
Variable B: accumulator zero

Using the equation below generate A from the given lines.

$$A = (A + B)\overline{B} + AB$$

26A Variable A: accumulator positive
Variable B: accumulator zero

A	B	$A + B$	AB	B	A
0	0	0	0	0	0
0	1	1	0	1	0
1	0	1	0	0	1
1	1	1	1	1	1

($A + B$, AB, B: given; A: desired)

Proof that equation is correct

$$\begin{aligned} A &= (A + B)\overline{B} + AB \\ &= A\overline{B} + B\overline{B} + AB \\ &= A\overline{B} + AB \\ &= A \end{aligned}$$

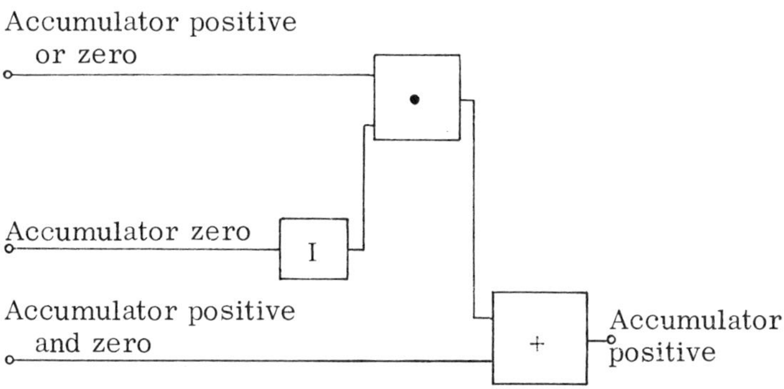

---

27 **Summary**

In this chapter, the translation of an English sentence into some form of a Boolean function was treated. Below are the most important points which were covered.

1. A phrase was defined to be a part of an English sentence which can be either true or false.

2. The problem must give and define all cases uniquely. This uniqueness is achieved by using the words "if and only if" (abbreviated iff).

3. Phrases fall into two categories, those which are set equal to dependent variables and those which are set equal to independent variables.

4. From the structure of the sentence, a truth table or equation can be found which defines a Boolean function.

5. Adding the word "not" in a phrase is equivalent to inverting the variable that phrase stands for.

6. A Boolean function can be found for a sentence even though the exact meaning of the individual phrase is not known.

7. When generating a circuit for a Boolean function which was derived from a sentence, the inputs and outputs can be labeled with the phrases instead of with the variables.

8. If the desired inputs to a circuit are not available they can often be generated from other related inputs.

PROBLEMS

1. Write an equation for the following sentence:

   Type out on the computer console typewriter when (iff) the computer has completed the problem, when there is a programming error, or when an illegal operation is specified.

2. Design a circuit for Prob. 1.

3. Design a circuit whose output is 1 whenever Mr. Jones goes to the movies (first write a single sentence for the problem, then treat as an ordinary problem).

Mr. Jones will go to the movies only if he can get a baby sitter.
Mr. Jones never goes to the movies if it rains, or on weekends.
Mr. Jones will only go if a Western is playing.

4. Write an equation to fit the following sentence:

   Store the content of the accumulator when (iff) a STORE instruction is given, at the end of an arithmetic instruction when no error occurred, or when the console typewriter is used.

5. In computer installations audible alarms are sometimes installed. Design a circuit which will feed such an alarm and sound it at the time indicated.

   Sound alarm when (iff) no power is present and the computer power switch is on.

   The following signals are available:
   Power present: variable A
   Computer off: variable B

6. Design a circuit whose output signal will be 1 when either input A or input B or both is 1. A and B are not available to the designer, but M and N are.

   M is 1 if A and B are 1; N is 1 if A or B but not both are 1 (this signal will be 1 if either of the two is 1, but 0 if both are 1).

7. A paper-tape reader is used to supply information to a computer. The paper tape shown in the diagram moves across reading station R where the hole pattern is translated into electrical voltage levels carried on four lines (one per track).

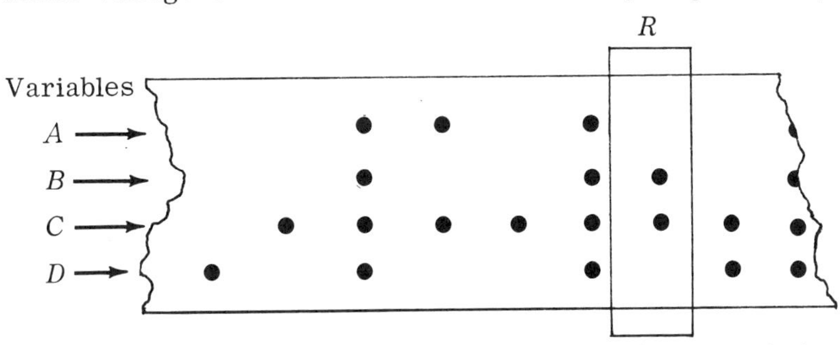

   All lines are normally 0 except when a pattern of holes is being sensed across the tape. When a hole is sensed, the associated line goes to the 1 level. All patterns of holes are allowed, except the pattern consisting of *none* of the four possible holes. This pattern is not used because one would not be able to distinguish between that pattern and the unpunched area between patterns.

(*a*) Write an algebraic equation for Z. Variable Z should be 1 only when a pattern is being sensed. Remember that every pattern must have at least one hole punched.

(*b*) If an incorrect pattern is punched on tape, it can be "erased" by repunching all four holes in that position. Whenever the computer senses the pattern in which all four holes are punched, it must treat that pattern as if it were unpunched tape. Write an equation for Y. Variable Y should be 1 only when the "erased" pattern is being sensed.

(*c*) Assume that the pattern chosen to represent the decimal digit 6 consists of holes in the B and C positions but not in the A and D positions. Write an equation for X. Variable X should be 1 only when the pattern for 6 is being sensed. Remember that there may be other patterns which have holes in the B and C positions.

8. Write the truth table for an expression whose value is 1 only when the variables A and B are different. Label the right column of the table "expression."

9. Write the truth table for an expression whose values is 1 only when an odd number of the variables A, B, and C are equal to 1. Zero should be considered an even number.

PART II

MINIMIZATION TECHNIQUES

In this part formal minimization techniques are introduced. These techniques are based on certain theorems developed in Part I. The minimization techniques which are introduced will produce answers in two levels, that is, the AND of variables ORed together (or the OR of variables ANDed together).

These techniques generate a minimum solution in accordance with the following priority:

1. Minimum number of terms

2. Minimum number of letters in each term

The number of inverters is not minimized.

Chapter 7

FORMS OF EXPRESSIONS

We have worked enough with Boolean algebra to know that any expression can be written in a number of equivalent forms. It is the purpose of this chapter to show some of the general forms an expression can take, and how we may change from one form to another. The names of the forms to be covered are the sum form, product form, standard sum form, and standard product form. In addition, binary and decimal notation for the standard sum form will be introduced.

1Q Basically, an expression can be written in either a *sum* form or a *product* form. An expression is in a *sum* form if it is made up of several smaller expressions ORed together. Examples for the *sum* form are:

$$A + B \qquad A + B\overline{C} \qquad A(B + C) + DE(G + M)$$

An expression can also be written in *product* form. The criterion for this form is that the expression is the AND of several smaller expressions. Examples of the *product* form are:

$$AB \qquad D(A + BC)(E + \overline{A}) \qquad (A + \overline{B})(D + \overline{E})$$

What form is the following expression in?

$$A\overline{B}(\overline{C} + D) + E\overline{G}$$

1A Sum form

This is so because it can be broken down into expressions which are ORed together. $A\overline{B}(\overline{C} + D)$ is ORed with $E\overline{G}$.

2Q What form is this expression in?

$$A(\overline{B}\overline{C} + D) + \overline{B}$$

2A Sum

3Q The following expression is in ______ form.

$$(A + B)(C + D)G$$

3A Product

The expression is the AND of other expressions.

---

4 In conventional algebra the word "factor" is usually associated with the *product*. For example, in the expression "A times B," the A and B are referred to as factors. The same holds true in Boolean algebra. *Factors* are those expressions or variables which are ANDed together to give an expression in the *product* form.

In conventional algebra the word "term" is usually associated with the *sum*, as in A plus B. In Boolean algebra the *terms* refer to the parts of an expression which are ORed together, giving rise to the *sum* form.

In the *sum* form:	*terms* are ORed
In the *product* form:	*factors* are ANDed

5Q If an expression is in the product form, the parts which are ANDed together are referred to as (terms) (factors).

5A Factors

The word "factor" comes from conventional algebra and is usually associated with multiplication, which is closely related to the AND operation in Boolean algebra.

6Q If an expression is in the sum form, the parts that are ORed together are referred to as (terms)(factors).

6A Terms

7Q **Example**

Let us look at an expression and its inverse.

Expression: $(A + B)(C + D)$
Inverse of expression: $\overline{A}\overline{B} + \overline{C}\overline{D}$

If an expression is in *product* form, its inverse is in ______ form.

7A Sum

---

8 Recall the two forms of De Morgan's theorem.

$$\overline{A + B + C + \cdots} = \overline{A}\,\overline{B}\,\overline{C} \cdots \qquad \text{and}$$
$$\overline{ABC \cdots} = \overline{A} + \overline{B} + \overline{C} + \cdots$$

In one case the expression is in *sum form* $(A + B + C)$, and after inverting the expression and carrying the inversion out, the result is in product form $(\overline{A}\overline{B}\overline{C})$. The opposite holds true if the original expression is in *product* form. In this case the final answer is in *sum* form. As all expressions fall into either one of these two types of expression this general statement can be made: Inversion changes expressions from *sum* form to *product* form and vice versa. This observation will be used later in the chapter. First, however, we shall investigate some of the possible *sum* forms an expression can take on.

---

9 In this frame an expression is changed into different sum forms by use of already known theorems.

$$X = AB\overline{C} + (D + \overline{D})C$$

Realizing that $(D + \overline{D}) = 1$, the equation can be rewritten as

$$X = AB\overline{C} + C$$

Using another theorem which states that $A\overline{B} + B = A + B$, we can again rewrite the equation:

$$X = AB + C$$

Returning to the original form, the second term can be multiplied out.

$$X = AB\overline{C} + CD + C\overline{D}$$

With little effort the original equation has been changed into several forms. In this relatively simple example one expression takes on several forms and it is impossible to establish their equality by inspection. However, there are forms of every expression that are unique. These forms are the "standard forms."

10Q The first standard form to be discussed is the *standard sum form.*

Definition

An expression is in standard sum form if each term contains every variable (in either the true or the inverted form) and all variables are ANDed together within each term.

Example

Assume that X is a function of A, B, C, and D.

$$X = \overline{A}\overline{B}C + \overline{A}BCD$$

The expression for X is not in a standard sum form because in the first term D does not appear in either its true form or its inverted form.

$$X = \overline{A}\overline{B}CD + \overline{A}\overline{B}C\overline{D} + \overline{A}BCD$$

The expression for X is in standard sum form. Which of the following expression is in standard sum form? (Assume that all expressions are part of a four-variable problem with the variables A, B, C, and D.)

(1) $A\overline{B}\overline{C}\overline{D} + \overline{A}\overline{B}CD + \overline{A}B$

(2) $\overline{C}D + A\overline{B}C\overline{D}$

(3) $A\overline{B}\overline{C}D + \overline{A}BCD + ABC\overline{D}$

(4) $(A + B)CD + \overline{A}\overline{B}(C + D)$

10A Only 3

In expression 1, the last term is missing two variables. In expression 2, the first term lacks variables A and B. In expression 4, the terms are more complex, not just the AND of the variables.

11Q Assume that each expression is a function of the variables A, B, and C. Which are in standard sum form?

(1) $A\overline{B} + C$

(2) $\overline{A}BC + A\overline{B}C$

(3) $AB + \overline{B}\overline{C}$

(4) $A\overline{B}C + ABC$

(5) $(A + B + C)(\overline{A} + \overline{B} + \overline{C})$

(6) $\overline{A}B + A\overline{B} + C$

11A 2 and 4

In expression 1 the first term does not contain a C and the last term does not contain the variables A and B. In expression 3 the first term does not contain a C, the second term does not include A. Expression 5 is in product form. In expression 6 the first two terms lack a C, the last term lacks both A and B.

---

12 In Chap. 4 the truth table was used to show the equivalency of two expressions. We learned to plot on the table but not how to read it. Another use for the table is to place an expression on the table and read it in standard sum form.

As you recall, expressions were plotted on the truth table by finding all the rows which are equal to 1 and setting the rest equal to 0. When reading the table the opposite process is followed:

1. Locate the first row in the table for which the expression is 1.
2. Translate the row found in step 1 into one term of the answer by first reading the row as a string of 0's and 1's, then replacing these binary digits with variables in the same order in which they appear on the top of the table, and finally, inverting the variables which corresponded to 0's.
3. Search for the next row in the table for which the expression is 1 and repeat step 2. If there are no more 1's, go to step 4.
4. OR all terms found.

Example

A	B	C	X
0	0	0	0
0	0	1	0
0	1	0	1
0	1	1	1
1	0	0	0
1	0	1	0
1	1	0	0
1	1	1	1

Step 1: Choose 3rd row
Step 2: Read 010, translate to $\overline{A}B\overline{C}$
Step 3: Choose 4th row
Step 2: Read 011, translate to $\overline{A}BC$
Step 3: Choose last row
Step 2: Read 111, translate to ABC
Step 3: No more
Step 4: $X = \overline{A}B\overline{C} + \overline{A}BC + ABC$

13Q Read Y from the table.

A	B	C	Y
0	0	0	0
0	0	1	0
0	1	0	0
0	1	1	1
1	0	0	0
1	0	1	1
1	1	0	0
1	1	1	1

$Y =$ ____________________

13A

A	B	C	Y
0	0	0	0
0	0	1	0
0	1	0	0
0	1	1	1
1	0	0	0
1	0	1	1
1	1	0	0
1	1	1	1

$Y = \overline{A}BC + A\overline{B}C + ABC$

---

14 The preceding problems demonstrate that the answer is always in a sum form and that each term contains each variable. This, however, is the definition of the *standard sum form* of an expression. Thus we can formulate the following rule.

Rule: When reading a function from the truth table it is in *standard sum form.*

15Q Read the table below to obtain an expression for X in standard sum form.

A	B	X
0	0	0
0	1	1
1	0	1
1	1	1

$X =$ ______________________

15A $X = \overline{A}B + A\overline{B} + AB$

16Q Note that in the answer above the three terms each contain the variables A and B. Thus the form given above is the standard sum form of the expression. Convert the expression $\overline{A}B + A\overline{B}C$ into standard sum form by (1) plotting it on the table, and (2) reading it from the table.

A	B	C	$\overline{A}B$	$A\overline{B}C$	$\overline{A}B + A\overline{B}C$
0	0	0			
0	0	1			
0	1	0			
0	1	1			
1	0	0			
1	0	1			
1	1	0			
1	1	1			

$\overline{A}B + A\overline{B}C =$ ______________________

16A $\overline{A}B\overline{C} + \overline{A}BC + A\overline{B}C$

A	B	C	$\overline{A}B$	$A\overline{B}C$	$\overline{A}B + A\overline{B}C$
0	0	0	0	0	0
0	0	1	0	0	0
0	1	0	1	0	1
0	1	1	1	0	1
1	0	0	0	0	0
1	0	1	0	1	1
1	1	0	0	0	0
1	1	1	0	0	0

17Q The truth table was used to change an expression into a standard sum form. Suppose that X was given and $\overline{X}$ was to be found in standard sum form. The first step is to plot X on the table. This step has been done for you for $X = \overline{A}B + A\overline{B}$. Next find $\overline{X}$ on the table.

A	B	X	$\overline{X}$
0	0	0	
0	1	1	
1	0	1	
1	1	0	

17A

A	B	X	$\overline{X}$
0	0	0	1
0	1	1	0
1	0	1	0
1	1	0	1

18Q The final step is to read $\overline{X}$ in standard sum form from the table. $\overline{X}$ = ______________.

18A $\overline{X} = \overline{A}\overline{B} + AB$

19 **Review**

In the last few frames the standard sum form was introduced:

1. The standard sum form can be written for the true form of the function.

2. Or, it can be written for the inverse of the function.

Both forms can be written from the truth table.

20Q In frame 8 it was shown that applying De Morgan's theorem will change an expression from sum form to product form. The new form obtained by the inversion is called the *standard product form*. The following equation is in standard form; invert it.

$$X = \overline{A}\overline{B} + \overline{A}B + A\overline{B}$$

20A $\overline{X} = (A + B)(A + \overline{B})(\overline{A} + B)$

---

21 This form of $\overline{X}$ is called the standard product form. Its definition is very similar to that of the standard sum form.

Definition

An expression is in standard product form if each factor contains every variable (in either the true or inverted form) and all variables are ORed together in each factor.

Compare this definition to the one for the standard sum in frame 10, page 115. The *standard product form* of an expression can be obtained by:

1. Writing the inverse of the desired expression in standard sum form.

2. Inverting the result obtained in step 1.

22Q Here is a review of the steps used to obtain the standard product form of X.

1. Plot X on a truth table.

2. On the table find $\overline{X}$.

3. Read $\overline{X}$ from the table. ($\overline{X}$ is now in standard sum form.)

4. Algebraically invert $\overline{X}$ to obtain X.

Put X in standard product form: $X = \overline{A}\overline{B} + AB$.

22A

$X = \overline{A}\overline{B} + AB$

$\overline{X} = \overline{A}B + A\overline{B}$

(from table)

A	B	X	$\overline{X}$
0	0	1	0
0	1	0	1
1	0	0	1
1	1	1	0

$X = (A + \overline{B})(\overline{A} + B)$ [answer]

23Q The values of Y as a function of A, B, and C are plotted on the truth table below. Find the *standard product form* of this function (*Note:* standard product, not standard sum).

A	B	C	Y
0	0	0	1
0	0	1	1
0	1	0	1
0	1	1	0
1	0	0	1
1	0	1	1
1	1	0	0
1	1	1	0

$Y =$ ____________________

23A $Y = (A + \overline{B} + \overline{C})(\overline{A} + \overline{B} + C)(\overline{A} + \overline{B} + \overline{C})$

The starting point is the inverse of Y in standard sum form:

$$\overline{Y} = \overline{A}BC + AB\overline{C} + ABC$$

Inverting this equation yields the answer.

---

24 **Review**

There are two standard forms:

1. Standard sum form

2. Standard product form

Both the function and its inverse can be written in either of the above forms. The standard product form for a function is obtained by inverting the standard sum form of the inverse of the function.

25Q Let us turn our attention to a method that will produce the *standard sum form* without the use of the truth table. Suppose $X = ABC + \overline{A}\overline{B}$ is to be put into *standard sum form*. The first term includes all variables; therefore it can remain unchanged. The second term, however, does not include the C. Let us multiply the last term by 1, which does not change its value, but let us write this 1 as $(C + \overline{C})$. C was chosen because this is the missing variable. This process of adding variables to terms is called *expanding*. Multiply out

$$X = ABC + \overline{A}\overline{B}(C + \overline{C}) = \_\_\_\_\_\_\_\_\_\_\_\_\_\_\_\_$$

25A $X = ABC + \overline{A}\overline{B}C + \overline{A}\overline{B}\overline{C}$

26Q Assuming that X is a function of the variables A, B, C, and D, expand the expression to the standard sum form:

$$X = \overline{A}\overline{B}C + A\overline{B}D$$

26A $X = \overline{A}\overline{B}C + A\overline{B}D$

$X = \overline{A}\overline{B}C(D + \overline{D}) + A\overline{B}D(C + \overline{C})$

$X = \overline{A}\overline{B}CD + \overline{A}\overline{B}C\overline{D} + A\overline{B}CD + A\overline{B}\overline{C}D$ [answer]

27Q In some problems the expansion process produces several identical terms in the expression. By applying the theorem $A + A = A$ all but one of these identical terms can be eliminated. Expand this expression to the standard sum form and apply the theorem.

$$X = ABC + ACD + ABD$$

27A

$$\begin{aligned} X &= ABC + ACD + ABD \\ &= ABC(D + \overline{D}) + ACD(B + \overline{B}) + ABD(C + \overline{C}) \\ &= \underline{ABCD} + ABC\overline{D} + \underline{ABCD} + A\overline{B}CD + \underline{ABCD} + AB\overline{C}D \end{aligned}$$

$$X = ABCD + ABC\overline{D} + A\overline{B}CD + AB\overline{C}D \qquad \text{[answer]}$$

28Q If several variables are missing from one term, all are added by repeating the described step.

X is a function of A, B, and C

$X = A + BC$

In standard sum form:

$X =$ ________________________

28A

$$X = A(B + \overline{B}) + (A + \overline{A})BC$$

$$\begin{aligned} &= AB + A\overline{B} + ABC + \overline{A}BC \\ &= AB(C + \overline{C}) + A\overline{B}(C + \overline{C}) + ABC + \overline{A}BC \\ &= ABC + AB\overline{C} + A\overline{B}C + A\overline{B}\,\overline{C} + ABC + \overline{A}BC \end{aligned}$$

There are two terms ABC, so one is dropped.

$$X = ABC + AB\overline{C} + A\overline{B}C + A\overline{B}\,\overline{C} + \overline{A}BC \qquad \text{[answer]}$$

---

29 Two methods have been discussed to obtain the standard sum form of an expression:

1. Truth table

2. Algebraic expansion

Both methods will work for any size problem, but the truth table is usually used only for small problems because for large problems it becomes too cumbersome. Algebra is usually used for problems with a larger number of variables. In the following frames a numerical notation for the standard sum form is introduced. First, let us quickly review the binary number system.

---

30 This frame is intended as a review of the binary number system. If this is your first exposure or you find the frame confusing, go to Appendix B and then return to this frame.

As in the decimal system the right-most digit is the units digit (assuming no binary point is in the number). Each digit position has a fixed multiplier or weight which is dependent only on its relative position to the units position. The sum of all weights times the number associated with that weight gives the value of the number. The weights of the digits starting at the units position and working *left* are 1, 2, 4, 8, 16, etc. (They are the powers of 2 starting with 2^0.)

Example

$$\begin{aligned}10110 &= 1 \times 2^4 + 0 \times 2^3 + 1 \times 2^2 + 1 \times 2^1 + 0 \times 2^0 \\ &= 1 \times 16 + 0 \times 8 \quad + 1 \times 4 \quad + 1 \times 2 \quad + 0 \times 1 \\ &= 16 \quad + \quad 0 \quad + \quad 4 \quad + \quad 2 \quad + \quad 0 \\ &= 22\end{aligned}$$

10110 (binary) = 22 (decimal)

If this is not a sufficient review, read Appendix B and then return to this frame.

31Q Reading the table the first term would be $\overline{A}B\overline{C}$. But instead of using the variables we will now use the binary entries of the variables. For $\overline{A}B\overline{C}$ this would be 010.

A	B	C	X
0	0	0	0
0	0	1	0
0	1	0	1
0	1	1	0
1	0	0	1
1	0	1	0
1	1	0	0
1	1	1	0

The second term is: in variables,______; in binary notation, _____.

31A $A\bar{B}\bar{C}$; 100

The second row for which X is equal to 1 is the row in which $A = 1$, $B = 0$, and $C = 0$. The binary number describing this row is merely the combination of the three binary digits 100.

32Q Summarizing the last two frames:

Binary equivalent of $\bar{A}BC$ is 011
Binary equivalent of $A\bar{B}\bar{C}$ is 100

To keep track of the variables and their order we will make use of the function notation introduced in Chap. 4, frame 21, page 59. If X is a function of A, B, and C and $X = \bar{A}BC + A\bar{B}\bar{C}$, the equation can be rewritten in terms of binary notation:

$X(A, B, C) =$ ______ + ______

32A $X(A, B, C) = 011 + 100$

33Q The equation in frame 32A contains *all* the information the original equation contained. The Σ indicates a summation in conventional algebra; here it will denote the OR of several terms. For example,

$$Y(A, B, C) = 000 + 111$$

Y can also be written as

$$Y(A, B, C) = \Sigma(000, 111)$$

Rewriting X from frame 32A in this notation gives

$X(A, B, C) =$ __________

33A $X(A, B, C) = \Sigma(011, 100)$

34Q Z is plotted on the table. Write it in the desired format.

A	B	C	Z
0	0	0	0
0	0	1	1
0	1	0	1
0	1	1	1
1	0	0	0
1	0	1	0
1	1	0	0
1	1	1	1

$Z(A, B, C) = \Sigma($__________$)$

34A $Z(A, B, C) = \Sigma(001, 010, 011, 111)$

35Q Summary

It is important to note that, in this notation, each term has the same length as each of the other terms because each term contains each variable and each binary digit stands for the state of one variable, 1 for the true state, 0 for the inverted state. Thus this form is a standard sum form where the variables were replaced by binary digits. Of course it is important not to interchange these digits, because only their positions within the term define the variables they represent. It is easy to go from this notation to the table. Try plotting X.

$$X(A, B) = \Sigma(10, 11)$$

A	B	X
0	0	
0	1	
1	0	
1	1	

35A

A	B	X
0	0	0
0	1	0
1	0	1
1	1	1

$$X(A, B) = \Sigma(10, 11)$$

36Q

Plot the function Z on the table.

$$Z(A, B, C) = \Sigma(000, 001, 010, 011, 101, 111)$$

A	B	C	Z
0	0	0	
0	0	1	
0	1	0	
0	1	1	
1	0	0	
1	0	1	
1	1	0	
1	1	1	

36A

A	B	C	Z
0	0	0	1
0	0	1	1
0	1	0	1
0	1	1	1
1	0	0	0
1	0	1	1
1	1	0	0
1	1	1	1

37Q The problem in frame 36Q was given in terms of binary numbers. Each number was three digits long and uniquely represented one of the rows on the truth table. Each group of three binary digits can be considered a binary number and can then be translated into a decimal number. Translate the function Z into decimal notation by:

1. Considering each term a three-digit binary number.

2. Translating each binary number into one decimal number. (You may use the Decimal to Binary Conversion Table at the end of the book.)

$$Z(A, B, C) = \Sigma(000, 001, 010, 011, 101, 111)$$
$$Z(A, B, C) = \Sigma(\_\_, \_\_, \_\_, \_\_, \_\_, \_\_)$$

37A $Z(A, B, C) = \Sigma(0, 1, 2, 3, 5, 7)$

The first term 000 is equivalent to decimal 0, the second term 001 is equivalent to decimal 1, the third term 010 is equivalent to decimal 2, etc.

38Q Starting with an expression in decimal notation, there are two steps involved in plotting it on the table:

1. Translate each decimal number to the binary equivalent.

2. Make $Y = 1$ for each row corresponding to each binary equivalent found in step 1.

Note: Each decimal digit corresponds to a 1 in the Y column, which in turn corresponds to a one term in the standard sum form of Y.

Plot Y on the table:

$Y(A, B, C) = \Sigma(0, 2, 7)$

A	B	C	Y
0	0	0	
0	0	1	
0	1	0	
0	1	1	
1	0	0	
1	0	1	
1	1	0	
1	1	1	

38A

A	B	C	Y
0	0	0	1
0	0	1	0
0	1	0	1
0	1	1	0
1	0	0	0
1	0	1	0
1	1	0	0
1	1	1	1

$Y(A, B, C) = \Sigma(0, 2, 7)$
$Y(A, B, C) = \Sigma(000, 010, 111)$

39

Summary

In this chapter, the following forms of expression have been introduced:

1. General sum forms
2. General product forms
3. Standard sum form
4. Standard product form
5. Sum and product form of the inverse
6. Binary notation
7. Decimal notation

Transformation from one form to any of the other forms was discussed.

PROBLEMS

1. Is $X = Y$?
   (a) $X = AB + AC + BC$
   (b) $Y = \bar{A}BC + A\bar{B}C + AB$

2. Is $U = V$?
   (a) $U = \bar{A}\bar{B}\bar{C} + \bar{A}BD + ABC + A\bar{B}\bar{D}$
   (b) $V = \bar{B}\bar{C}\bar{D} + \bar{A}\bar{C}D + BCD + AC\bar{D}$

3. Which of the following four expressions are equivalent?
   (a) $\bar{B}\bar{D} + \bar{A}\bar{B}D + BCD + A\bar{B}$
   (b) $\bar{A}B\bar{C}D + ACD + \bar{B}\bar{D} + BCD$
   (c) $\Sigma(0, 2, 5, 7, 8, 9, 10, 11, 15)$
   (d) $A\bar{B}\bar{C} + BCD + \bar{A}\bar{B}\bar{D} + A\bar{B}C + \bar{A}BD + A\bar{B}D$

4. Is $X = Y$?
   (a) $X = \bar{A}\bar{C} + \bar{A}B + A\bar{B}$
   (b) $Y = (\bar{A} + \bar{B})(A + B + \bar{C})$

5. $X = A\bar{B} + \bar{C}$.
   Write (a) X in some sum form
   (b) $\bar{X}$ in some product form
   (c) X in standard sum form
   (d) X in standard product form
   (e) $\bar{X}$ in standard sum form
   (f) $\bar{X}$ in standard product form
   (g) X in binary notation
   (h) X in decimal notation

6. $Y = \bar{A}\bar{B} + \bar{B}\bar{C}$
   Write (a) Y in some sum form
   (b) $\bar{Y}$ in some product form
   (c) Y in standard sum form
   (d) Y in standard product form
   (e) $\bar{Y}$ in standard sum form
   (f) $\bar{Y}$ in standard product form
   (g) Y in binary notation
   (h) Y in decimal notation

7. $X = A + B + C$. What standard form is X in?

Chapter 8

QUINE-McCLUSKEY METHOD

This chapter teaches the Quine-McCluskey method of minimizing a Boolean function. Using this method, a function in standard sum form is reduced to a set of terms, called *prime implicants*, from which as many variables as possible have been eliminated. The Quine-McCluskey method is shown to be a simplified procedure for repeated use of the theorem $AB + A\bar{B} = A$, but using binary notation, and with rules for limiting the number of necessary applications of the theorem. In Chap. 9, it will be shown how some of these prime implicants can be eliminated.

---

1 The Quine-McCluskey method assumes that the expression is in the binary notation for the standard sum form. The algerism itself does nothing more than repeatedly apply the theorem $AB + A\bar{B} = A$. Basically this means that terms are combined into new terms. The advantage of the algerism is that no guesswork is involved and that it finds *all* possible pairs of terms which can be combined. When terms cannot be further combined, this algerism has produced what will be defined as *prime implicants*. In Chap. 9 it will be shown that sometimes not all these prime implicants are needed. To select the proper set or sets of prime implicants the *prime-implicant chart* is used (Chap. 9).

2Q Multiplying out was covered in Chap. 4, page 58, frame 16. For example, $A(B + C)$ can be multiplied out to give $AB + AC$. We concluded that multiplying out in Boolean algebra follows the same rules as in conventional algebra.

The inverse operation, factoring, also follows the same set of rules in both conventional and Boolean algebra.

For example, B can be factored out of the Boolean expression $ABC + B\overline{C}$:

$$ABC + B\overline{C} = B(AC + \overline{C})$$

Note that C and $\overline{C}$ can *not* be factored out because C and $\overline{C}$ are essentially different. Factor the following expression: $AB + A\overline{B} =$ ______________.

2A $AB + A\overline{B} = A(B + \overline{B})$

3Q You may note that the expression in parentheses in the answer above reduces to 1. Thus the following theorem can be formulated.

Theorem

$$AB + A\overline{B} = A$$

Note what this theorem does. It combines two terms which differ by *one* bar over *one* variable; or the same variable appears in both terms, once in its true form, once in its inverted form. The theorem does ***not*** apply to these pairs of terms:

$\overline{A}\overline{B} + AB$		cannot be factored
$A\overline{B}\overline{C} + ABC$	$= A(\overline{B}\overline{C} + BC)$	expression in parentheses is not of the proper form
$\overline{A}BC + AB$	$= B(A + \overline{A}C)$	expression in parentheses is not of the proper form

Using this theorem, which of the following pairs of terms can be combined in two steps:

1. Factoring out all but one variable

2. Replacing the unfactored expression by 1

1. $\bar{A}BC + ABCD$

2. $\bar{A}B\bar{C} + AB\bar{C}$

3. $\bar{A}\bar{B}C + ABC$

3A Only expression 2; the result is $B\bar{C}$

Expression 1 can be factored to:

$$\bar{A}BC + ABCD = BC(\bar{A} + AD)$$

The expression in parentheses is not equal to 1. As for expression 3, the theorem does not apply. Factoring would give:

$$\bar{A}\bar{B}C + ABC = C(\bar{A}\bar{B} + AB)$$
$$\bar{A}\bar{B} + AB \neq 1$$

Therefore the theorem does not apply.

4Q Using the theorem $AB + A\bar{B} = A$, the expression

$$X = \bar{A}B\bar{C} + \bar{A}BC + AB\bar{D} + ABD$$

can be simplified to

$X =$ ______________

4A $X = B$

$\bar{A}B\bar{C} + \bar{A}BC$ can be simplified to $\bar{A}B$. $AB\bar{D} + ABD$ can be simplified to AB. $\bar{A}B$ and AB combine to give the answer, B.

5Q In Chap. 7 the binary notation for expressions was introduced. If the expression is given in this form, it can be simplified directly without having to convert to variables.

In the binary number form, digits are eliminated in the process of simplification. Thus the new term is shorter by one binary digit. But as the position of each digit determines the variable the digit stands for, the position within the number, and therefore the length of each term, must not be changed. To avoid confusion and make all terms of equal length, the variables which have been eliminated are replaced by dashes. (A variable in the true form is represented by a 1; a variable in the inverted form is represented by a 0; the absence of a variable from the term is indicated by the symbol -.)

Example

Assume that the problem has the variables A, B, and C. Then the term

$A\bar{B}C$ is written as 101
$\bar{A}\bar{B}\bar{C}$ is written as 000
$A\bar{B}$ is written as 10-

Complete the following list:

$\bar{A}B\bar{C}$ is written as ______
A is written as ______

5A $\bar{A}B\bar{C}$ is written as 010; A is written as 1--

---

6 Let us follow the process of eliminating a variable in both conventional and binary notation. For example, combine $AB + A\bar{B}$; the answer is A. Combine 11 + 10; the answer is 1-. *Note:* The symbol + is still used to indicate the OR operation, even though it is associated with numbers.

---

7 Let us apply the theorem $AB + A\bar{B} = A$ to two terms with a larger number of variables. A stands for all variables common to the terms, B for the rest.

Example 1

$$\underbrace{CDE}\,\underbrace{F} + \underbrace{CDE}\,\underbrace{\bar{F}} = ?$$
$$A \quad B + \quad A \quad \bar{B} = A$$
$$CDEF + CDE\bar{F} = CDE$$

Example 2

Suppose, however, that more than one variable is contained in B:

$$\underbrace{CD}\,\underbrace{\bar{E}F} + \underbrace{CD}E\bar{F}$$
$$A \quad B \; + \; A$$

If $\bar{E}F = B$, is $\bar{B} = E\bar{F}$? No. Therefore the theorem does not apply. In summary we can say that unless two terms differ by one bar over one variable they cannot be combined. Here is a comparison of the two examples given in this frame with their binary representation:

Example 1

$$CDEF + CDE\bar{F} = CDE$$
$$\underbrace{111}\,\underbrace{1} + \underbrace{111}\,\underbrace{0} = 111\text{-}$$

Example 2

$CD\bar{E}F + CDE\bar{F}$ (do not combine)
$1101 + 1110$ (do not combine)

---

8 It has thus been established that if two terms can be combined they must differ in only one digit position. Suppose we were to count the number of 1's in each of the two terms that can be combined. We would find that one of the terms would have one more 1 than the other. Reasoning this out is rather simple: The first term has M 1's. The second term is the same as the first term except

that one of the digits has been changed from a 0 to a 1 or from a 1 to a 0. Thus the second term must have either $(M - 1)$ or $(M + 1)$ 1's. The reasoning indicates that the number of 1's in a given term is very important; thus it will be given a special name: *index* of the term.

Definition

The *index of a term* is the number of 1's in that term. The following rule can be established as a necessary but not sufficient condition for combining terms.

Rule: The indexes of two terms must differ by exactly one.

Example

The index of the term 10011 is 3. The index of the term 11000 is 2.

9Q We are now ready to start the Quine-McCluskey method. Assuming that the problem has been given in the binary notation of the standard sum form (any problem can be translated into this form), the first step is to sort the terms into groups of constant indexes:

$$X(A, B, C) = \Sigma(000, 010, 011, 100, 110, 111)$$

Compute the index for each term and classify the terms according to their indexes.

9A

Index 0:	000	Index 2:	011
			110
Index 1:	010		
	100	Index 3:	111

---

10 In this problem the terms can be sorted into four groups of constant index. It is customary to list all terms in one column and to separate the groups of constant index by horizontal lines. Rewriting the example from frame 9A gives

000

010
100

011
110

111

11Q Rewrite the expression for Y in tabular form in preparation for the Quine-McCluskey method. Perform these steps:

1. Find the index of each of the terms.
2. Sort the terms by their indexes.
3. Place them in tabular form, separating groups of constant index by horizontal lines.

$$Y(A, B, C, D) = \Sigma(0000, 0010, 0011, 0100, 0101, 0111, 1010, 1110)$$

11A

0000

0010
0100

0011
0101
1010

0111
1110

12Q When using the prime-implicant chart (Chap. 9) it will become necessary to identify each of the terms. To ensure unique identification each term is identified by its decimal equivalent. The decimal equivalents of the first three terms are entered in the table. Enter the remaining decimal equivalents. (You may wish to use the conversion table at the end of the book.)

0	0000
2	0010
4	0100
	0011
	0101
	1010
	0111
	1110

12A

0	0000
2	0010
4	0100
3	0011
5	0101
10	1010
7	0111
14	1110

13 Having written the expression in the correct format, terms that can be combined are combined. Let us now see how terms can be combined.

1. We know that only pairs of terms can be considered. This conclusion is drawn directly from the theorem $AB + A\bar{B} = A$.

2. Pairs need only be considered if their index differs by one.

Rule: Two binary terms can be combined if they differ in one binary digit. This digit is replaced by a dash (-) to ensure proper alignment.

This rule for combining binary terms is essentially a restatement of the theorem $A\bar{B} + AB = A$. For example,

Given:	$\bar{A}B\bar{C}D$	or in binary	0101
	$\bar{A}BCD$		0111
Combine to:	$\bar{A}B\ D$	combine to	01-1

14Q In this example, two index groups are given. Compare the entry in the top group with all entries in the bottom group: 1, 3; 1, 5; 1, 6; etc. Which pairs can be combined?

1	0001
3	0011
5	0101
6	0110
10	1010

14A 1, 3 and 1, 5

1, 6 cannot combine; they differ in three digits. The same is true for 1, 10.

15Q Again the problem has two index groups. To find *all* pairs of terms which can be combined, *all* entries in the upper group have to be compared to *all* entries in the lower group. The customary approach is to compare the top entry (1 0001) with all entries in the lower group (3 0011), (5 0101), and (12 1100). Then the second entry in the top group (2 0010) is compared with all entries in the lower group. If there is a third term in the top group it is used next. This exhaustive comparison of terms assures us that "everything" has been tried. But it also is a common cause for omissions. Find all pairs of terms which can be combined.

1	0001
2	0010
3	0011
5	0101
12	1100

15A 1, 3; 1, 5; 2, 3; or in list form:

1	0001	1	0001	2	0010
3	0011	5	0101	3	0011

16Q The pairs of terms found in the last frame can be combined to give the following new terms (remember to replace variables which were eliminated by -):

1, 3 ________
1, 5 ________
2, 3 ________

16A 1, 3 00-1
1, 5 0-01
2, 3 001-

The variables which were eliminated are replaced by a - to keep the position of the remaining variables the same.

---

17 How is the decimal identification handled? Both decimal numbers are now placed in front of the new term; thus

entries with one dash have two decimal numbers identifying them. The new terms are written in a new list, as shown in frame 18.

18 **Summary**

Assuming X is given in standard sum form, the entire problem is repeated here.

$$X(A, B, C, D) = \Sigma(0001, 0010, 0011, 0101, 1100)$$

1	0001	1, 3	00-1
2	0010	1, 5	0-01
3	0011	2, 3	001-
5	0101		
12	1100		

These steps are taken:

1. The index for each term is found.

2. Terms are sorted into groups of constant index.

3. Groups are sorted by indexes and written in one column, groups with small index on top, groups with larger index on bottom.

4. Decimal equivalent is found for each term and entered.

5. Entries in adjacent groups (index differing by 1) are compared; some pairs are combined.

6. Decimal numbers are carried along into a new list.

The original column of numbers is often referred to as *first list*, the list of new terms as *second list*. Higher-order lists will be generated later. The numbering for these lists is merely an extension.

19Q Write the expression for Z in the proper format for the Quine-McCluskey method and combine the terms which can be combined.

$$Z(A, B, C, D) = \Sigma(0000, 0010, 0100, 1000)$$

19A $Z(A, B, C, D) = \Sigma(0000, 0010, 0100, 1000)$

0	0000	0, 2	00-0
2	0010	0, 4	0-00
4	0100	0, 8	-000
8	1000		

20Q It is important that you keep in mind what these new binary terms stand for. Assuming that the problem above is written for the variables A, B, C, and D, the three terms in the second list written in terms of the variables are_______, _______, and _______.

20A $\bar{A}\bar{B}\bar{D}$; $\bar{A}\bar{C}\bar{D}$; $\bar{B}\bar{C}\bar{D}$

Each of the 0's changes into the inverted form of the variable. The dash means the variable has been eliminated. Thus the term 00-0 changes to $\bar{A}\bar{B}\bar{D}$. There is no C in this term because the dash indicates that it has been eliminated. If you want to check yourself refer back to frame 13, page 138.

21Q The following example is given in standard sum form. Translate it into binary notation and go through the steps of the method explained thus far.

$$V = \bar{A}B\bar{C}D + A\bar{B}\bar{C}D + \bar{A}\bar{B}\bar{C}D + \bar{A}B\bar{C}\bar{D} + AB\bar{C}\bar{D}$$

21A

$$V = \bar{A}B\bar{C}D + A\bar{B}\bar{C}D + \bar{A}\bar{B}\bar{C}D + \bar{A}B\bar{C}\bar{D} + AB\bar{C}\bar{D}$$

$$V(A, B, C, D) = \Sigma(0101,\ 1001,\ 0001,\ 0100,\ 1100)$$

Index: 2 2 1 1 2

1	0001	1, 5	0-01
4	0100	1, 9	-001
5	0101	4, 5	010-
9	1001	4, 12	-100
12	1100		

22Q When two terms are combined in Boolean algebra, they are replaced by a new term. To indicate which terms have been combined and thus replaced, and which ones have not, it is customary to check the terms used in the combining process. Even though one term may be used as often as needed, it is still checked only once. In this example, 1 and 5 are checked because they were used to form 1, 5 0-01.

1	0001✓	1, 5	0-01
4	0100	1, 9	-001
5	0101✓	4, 5	010-
9	1001	4, 12	-100
12	1100		

What other terms should be checked?

22A All terms in the first list:

1	0001✓	1, 5	0-01
4	0100✓	1, 9	-001
5	0101✓	4, 5	010-
9	1001✓	4, 12	-100
12	1100✓		

23Q From now on the checking-off of terms will be part of combining. In this problem there will be three index groups. To obtain all *possible* combinations we must compare all the members of the first group with all the members of the second group, and all members of the second group with all the members of the third group. (*Note:* The first and third group do not have to be compared because their indexes differ by more than 1.) Combine *all* possible terms, write a second list.

0	000
1	001
2	010
3	011
5	101

23A

0	000✓	0, 1	00-
1	001✓	0, 2	0-0
2	010✓	1, 3	0-1
3	011✓	1, 5	-01
5	101✓	2, 3	01-

24Q The new list (or second list) will be treated much like the original one. To use the notion of index again we will define the index to be the number of 1's, not counting the dashes as 1's. The first step is to sort the entries by their index. But they are already sorted, because as we work down the first list we are going from small indexes to large ones. As the index in the new list depends on the index of the terms generating it, the entries in the new list are automatically in order. Groups in the new list are separated by horizontal lines, exactly as in the original list. In the example above a horizontal line should be drawn between the entries ______ and ______.

24A 0, 2 0-0; 1, 3 0-1

The index of the first two entries is 0; the index for the remaining entries is 1.

25Q

1	001 ✓	1, 5	-01
2	010	5, 7	1-1
5	101 ✓		
7	111 ✓		

Once more recall what these binary numbers really mean. The original list was derived from the standard sum in which all terms are ORed. All this method has done thus far is show which terms can be combined; those which were combined were checked. In the example above, four terms are in the standard sum form. Three of them can be replaced by two new terms. Using the terms of both lists, copy all terms which have to be ORed to produce the simplified function.

25A

2	010
1, 5	-01
5, 7	1-1

The three remaining terms were checked off because they were replaced by the terms of the second list.

26Q From the list in frame 25A write an expression in terms of the three variables A, B, and C.

$X =$ ______ + ______ + ______

26A $X = \bar{A}B\bar{C} + \bar{B}C + AC$

A dash means the variable has been eliminated. A 1 means the variable is in its true form. A 0 means the variable is in its inverted form.

27Q Translate Y back to algebraic terms using the variables A, B, and C.

0	000 ✓	0, 1	00-
1	001 ✓	0, 4	-00
4	100 ✓	1, 5	-01
5	101 ✓	4, 5	10-

$Y =$ ______________

27A $Y = \overline{A}\overline{B} + \overline{B}\overline{C} + \overline{B}C + A\overline{B}$

All entries in the first list were checked off, thus none of them are needed in the expression for Y.

---

28 In the equation in frame 27A the terms $\overline{A}\overline{B} + A\overline{B}$ combine to $\overline{B}$ and the terms $\overline{B}\overline{C} + \overline{B}C$ also combine to $\overline{B}$. If this is so, it should be possible to do the same in the binary form; but note that all of these entries were in the second list and therefore had one dash each. To combine terms which appear in the second list, two conditions have to be satisfied:

1. The two terms must contain the same variables; i.e., all dashes *must* line up.

2. Only one digit may be different in the two terms.

---

29 Continuing the problem of frame 27 the binary notation is used to combine entries from the second list.

0, 1	00- ✓	0, 1, 4, 5	-0-
0, 4	-00 ✓	0, 1, 4, 5	-0-
1, 5	-01 ✓		
4, 5	10- ✓		

Combined entries are checked. Decimal numbers associated with a given term are rearranged in ascending order. Note the identical entries in the new list; this is a general property of all lists with more than onc dash. It is a useful check against omissions.

30Q This new problem has four entries in its first list. These entries will combine by pairs into a second list, from which we can derive a third list by going through the steps described in the previous frame. Perform all these steps generating a third list with two identical entries. Remember to place a check mark next to all entries which were combined.

0	000
1	001
4	100
5	101

30A

0	000 ✓	0, 1	00- ✓	0, 1, 4, 5	-0-
1	001 ✓	0, 4	-00 ✓	0, 1, 4, 5	-0-
4	100 ✓	1, 5	-01 ✓		
5	101 ✓	4, 5	10- ✓		

31Q In this example, generate all possible lists. Then find all remaining (unchecked) terms and write them by using the variables A, B, C, and D. Go slowly to avoid bookkeeping errors. If there exists a third list remember that entries come in identical pairs. Use this as a check.

2	0010
4	0100
3	0011
6	0110
9	1001
12	1100
7	0111

$X =$ ____________________

31A $X = A\overline{B}\overline{C}D + \overline{A}B\overline{D} + B\overline{C}\overline{D} + \overline{A}C$

2	0010 ✓	2, 3	001- ✓	2, 3, 6, 7	0-1-
4	0100 ✓	2, 6	0-10 ✓	2, 3, 6, 7	0-1-
3	0011 ✓	4, 6	01-0		
6	0110 ✓	4, 12	-100		
9	1001	3, 7	0-11 ✓		
12	1100 ✓	6, 7	011- ✓		
7	0111 ✓				

The unchecked terms are:

9	1001
4, 6	01-0
4, 12	-100
2, 3, 6, 7	0-1-

Using the variables:

$$X = A\overline{B}\overline{C}D + \overline{A}B\overline{D} + B\overline{C}\overline{D} + \overline{A}C$$

32 The Quine-McCluskey method was begun by combining terms in the original list to form a second list; then from the second list, a third list was developed. This process can be extended until the last list does not have two entries that can be combined to form an entry in a new list. When forming high-order lists, the same rules apply as for earlier lists (given in frame 28). It is important that *all* dashes line up. You should recall that in the third list all entries appear twice; one of them must be eliminated before going on. In the fourth list all entries appear in triplicate, and in the Nth list all entries appear in sets of $(N - 1)$. In all cases, all but one of the identical entries are eliminated before going on. This property lends itself very well to checking the results against omissions.

33 In this frame an entire problem is shown. Read through it and examine each step.

$$Z = \bar{A}BC + ABD + A\bar{B}\bar{C}D + AC\bar{D} + \bar{B}CD + \bar{A}\bar{B}C\bar{D}$$

Using the Quine-McCluskey method, reduce the expression for Z. Change it to standard sum form:

$$Z = \bar{A}\bar{B}C\bar{D} + \bar{A}\bar{B}CD + \bar{A}BC\bar{D} + \bar{A}BCD + A\bar{B}\bar{C}D + A\bar{B}C\bar{D} + A\bar{B}CD + AB\bar{C}D + ABC\bar{D} + ABCD$$

Compute indexes, list, combine, check, find identical terms, and eliminate all but one term by crossing them out.

2	0010 ✓	2, 3	001- ✓	2, 3, 6, 7	0-1- ✓	2, 3, 6, 7, 10, 11, 14, 15	--1-
3	0011 ✓	2, 6	0-10 ✓	2, 3, 10, 11	-01- ✓	~~2, 3, 6, 7, 10, 11, 14, 15~~	~~--1-~~
6	0110 ✓	2, 10	-010 ✓	~~2, 3, 6, 7~~	~~0-1-~~	~~2, 3, 6, 7, 10, 11, 14, 15~~	~~--1-~~
9	1001 ✓	3, 7	0-11 ✓	2, 6, 10, 14	--10 ✓		
10	1010 ✓	3, 11	-011 ✓	~~2, 3, 10, 11~~	~~-01-~~		
7	0111 ✓	6, 7	011- ✓	~~2, 6, 10, 14~~	~~--10~~		
11	1011 ✓	6, 14	-110 ✓	3, 7, 11, 15	--11 ✓		
13	1101 ✓	9, 11	10-1 ✓	~~3, 7, 11, 15~~	~~--11~~		
14	1110 ✓	9, 13	1-01 ✓	6, 7, 14, 15	-11- ✓		
15	1111 ✓	10, 11	101- ✓	~~6, 7, 14, 15~~	~~-11-~~		
		10, 14	1-10 ✓	9, 11, 13, 15	1--1		
		7, 15	-111 ✓	~~9, 11, 13, 15~~	~~1--1~~		
		11, 15	1-11 ✓	10, 11, 14, 15	1-1- ✓		
		13, 15	11-1 ✓	~~10, 11, 14, 15~~	~~1-1-~~		
		14, 15	111- ✓				

The only unchecked terms are 1--1 and --1-. Normally one intermediate step would be taken at this point to eliminate redundant unchecked terms. This step is discussed in Chap. 10. In this problem it has been omitted: $Z = AD + C$.

34Q The expression for Z, which was used as an example in frame 33, is repeated here in both the original and the reduced form.

$$Z = \bar{A}BC + ABD + A\bar{B}\bar{C}D + AC\bar{D} + BCD + \bar{A}\bar{B}C\bar{D}$$
$$Z = AD + C$$

Compute the following:

Cost of original form	*Cost of reduced form*	*Savings*
____transistors	____transistors	____transistors
____diodes	____diodes	____diodes

34A

Cost of original form	*Cost of reduced form*	*Savings*
4 transistors	0 transistors	4 transistors
26 diodes	4 diodes	22 diodes

In the original form all variables required inverting; thus 4 transistor inverters are required. Each input to an AND requires a diode (4 AND's with 3 inputs each plus 2 AND's with 4 inputs gives 20 diodes), and each input to an OR is a diode (6 diodes). Total for original form is 4 transistors and 26 diodes. The second form requires only 4 diodes and no transistors.

PROBLEMS

1. Simplify the expression for Z by:
   (*a*) Use of algebra and its theorems
   (*b*) Translating the expression for Z into standard sum form and then applying algebra and its theorems
   (*c*) Applying the Quine-McCluskey method; rewrite in terms of variables

$$Z = B\bar{C}\bar{D} + AB\bar{D} + \bar{A}BC\bar{D}$$

2. Simplify the function Y by:
   (a) Use of algebra and its theorems
   (b) Translating the expression for Y into standard sum form and then applying algebra and its theorems
   (c) Applying the Quine-McCluskey method; rewrite in terms of variables

$$Y = \bar{A}BD + BCD + A\bar{C}D + A\bar{B}CD$$

3. Using the Quine-McCluskey method, find an equivalent but cheaper circuit. How many diodes can be saved?

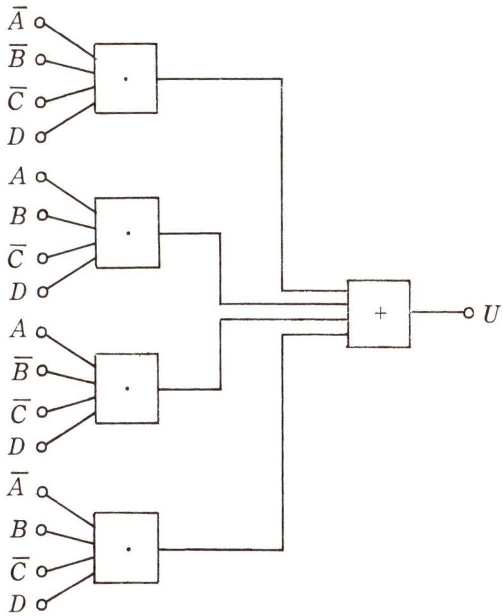

4. Find an equivalent but cheaper circuit using the Quine-McCluskey method. How many diodes can be saved?

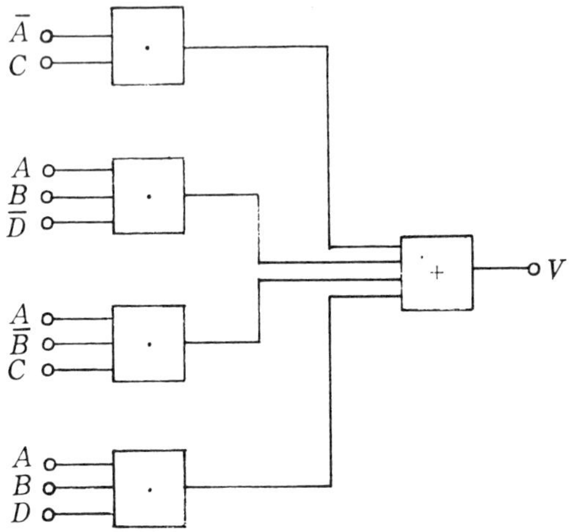

5. Consider two terms which can be combined. Why does their index necessarily differ by one?

6. Can the problem in frame 33 be simplified without using the Quine-McCluskey method but by using all the theorems?

Chapter 9

PRIME-IMPLICANT CHART

In this chapter, it is shown that not all of the terms (prime implicants) produced by the Quine-McCluskey method are needed, and a method is given (prime-implicant chart) for eliminating the unnecessary terms. The elimination is based upon the decimal numbers which show what terms had been combined to form the prime implicants.

1Q **Definition**

A *prime implicant* is a term which can not be combined any further using the Quine-McCluskey method.

In the example below, which terms are prime implicants?

$$X(A, B, C) = \overline{A}BC + A\overline{B}\overline{C} + AB\overline{C} + ABC$$
$$= 011 + 100 + 110 + 111$$

4	100 ✓	4, 6	1-0
3	011 ✓	3, 7	-11
6	110 ✓	6, 7	11-
7	111 ✓		

1A

4, 6 1-0
3, 7 -11
6, 7 11-

All other terms have been checked off, which means that they had been combined. They therefore are not prime implicants.

2Q Converting the prime implicants from the binary to the algebraic expressions, we find that the original expression (frame 1) has been reduced to $A\overline{C} + BC + AB$.

This set of prime implicants, however, is not the final answer to the problem. The reason is that some prime implicants may not be necessary. In the example under discussion, prime implicant AB is not necessary. Prove this by plotting both expressions on the truth table shown here.

A	B	C	$A\overline{C} + BC + AB$	$A\overline{C} + BC$
0	0	0		
0	0	1		
0	1	0		
0	1	1		
1	0	0		
1	0	1		
1	1	0		
1	1	1		

2A

A	B	C	$A\overline{C} + BC + AB$	$A\overline{C} + BC$
0	0	0	0	0
0	0	1	0	0
0	1	0	0	0
0	1	1	1	1
1	0	0	1	1
1	0	1	0	0
1	1	0	1	1
1	1	1	1	1

3Q The problem we now have to solve is that of finding out which prime implicants are not necessary. The decimal numbers used in the Quine-McCluskey method will be very helpful in this step, because they identify which of the standard sum terms combined to form each prime implicant. In the previous problem, the term 4, 6 1-0 means that the prime implicant 1-0 was obtained by combining standard sum terms 4 and 6. The term AB (6, 7 11-) is formed by combining _______________.

3A 6 and 7

4Q Since $A\bar{B}\bar{C}$ and $AB\bar{C}$ were combined to form $A\bar{C}$, we can say that $A\bar{C}$ *contains* both $A\bar{B}\bar{C}$ and $AB\bar{C}$. This is the same as saying that 1-0 contains 100 and 110 or as saying that 1-0 contains 4 and 6.

The prime implicants of the previous example are repeated below. It can easily be seen that one standard sum term may be contained in more than one prime implicant.

4, 6	1-0
3, 7	-11
6, 7	11-

Which standard sum terms (identified by their decimal number) appear in more than one prime implicant?

4A 6 and 7

6 is contained in 1-0 and 11-; 7 is contained in -11 and 11-.

5Q The reason a standard sum term may be contained in more than one prime implicant is that a term may be combined more than once in the Quine-McCluskey method. For example, 6 can be combined with 4 and also with 7.

4	100
3	011
6	110
7	111

In this example, 7 combines with ______.

5A 3 and 6

6Q Using the concept that a prime implicant contains a certain group of standard sum terms (specified by the decimal numbers), we are now ready to look for unnecessary prime implicants. The basic idea is that we *need just enough* prime implicants so that *all* the original standard sum terms (decimal numbers) are contained in at least one prime implicant of the final expression.

4, 6	1-0
3, 7	-11
6, 7	11-

The following prime implicants are *necessary:*

______________________.

6A

4, 6	1-0
3, 7	-11

The third prime implicant is not necessary because it contains terms 6 and 7, which are already contained in the first two prime implicants.

7 Summary

Prime implicants are obtained by combining standard sum terms using the Quine-McCluskey method. The decimal numbers in front of each prime implicant show the terms it contains. If *all* the standard sum terms contained in one prime implicant are included in other prime implicants, then that prime implicant is unncessary.

8Q The prime-implicant chart is a useful way of showing which terms are included in which prime implicants. By clearly displaying this information, the job of finding unnecessary prime implicants is simplified. To construct the prime-implicant chart *all* decimal numbers corresponding to the terms of the standard sum are listed horizontally, *all* prime implicants are listed vertically, and a grid is constructed. Let us try an example where the prime implicants are:

0, 4	-00
4, 5	10-
5, 7	1-1

	0	4	5	7
$\overline{B}\overline{C}$				
$A\overline{B}$				
AC				

Now all *inclusions* are indicated by X's. For $\overline{B}\overline{C}$, an X is placed under 0 and 4 because $\overline{B}\overline{C}$ *includes* 0 and 4. Complete the chart.

	0	4	5	7
$\overline{B}\overline{C}$	X	X		
$A\overline{B}$				
AC				

8A

	0	4	5	7
$\overline{B}\overline{C}$	X	X		
$A\overline{B}$		X	X	
AC			X	X

9Q Having plotted all prime implicants on the chart, we can now pick a set of them using the following rule.

Rule: Select the smallest set of prime implicants that includes all terms of the standard sum expression.

Two prime implicants include *all* terms of the standard sum expression. They are______and______.

9A $\bar{B}\bar{C}$ and AC

	0	4	5	7
$\bar{B}\bar{C}$	X	X		
$A\bar{B}$		X	X	
AC			X	X

$A\bar{B}$ is redundant.

10 Summary

The chart we have used lists the function in terms of the standard sum form along the top and the prime implicants of that function (found by the Quine-McCluskey method) along the side. X's are placed in each row under the standard sum terms contained in the prime implicant of that row. A sufficient set of prime implicants is selected to include at least one X in each column. Prime implicants not selected are redundant and are discarded. Since this chart deals exclusively with the selection of prime implicants, it is called the *prime-implicant chart.*

11Q A new prime-implicant chart is given here.

	1	4	5	9	13	14	15
$\bar{C}D$	X		X	X	X		
$\bar{A}B\bar{C}$		X	X				
ABD					X		X
ABC						X	X

In larger charts it is difficult to pick a set of prime implicants without a plan of attack. One approach is to first select the necessary prime implicants. A prime implicant is necessary if it is the only one which includes a term of the standard sum. Since every term has to be included, this prime implicant must be selected. These necessary prime implicants are easily found by looking for *columns* with a single X. The prime implicant which includes this single X is necessary. The first column contains one X. Therefore the prime implicant ______ is a necessary prime implicant.

11A $\bar{C}D$

Only $\bar{C}D$ has an X in the first column.

	1✓	4	5	9	13	14	15
$\bar{C}D$	X		X	X	X		
$\bar{A}B\bar{C}$		X	X				
ABD					X		X
ABC						X	X

12Q A check mark is used to indicate terms of the standard sum already included in the selected prime implicants. $\bar{C}D$ has been selected in order to include standard sum term 1. Aside from this term, prime implicant $\bar{C}D$ includes other standard sum terms. Check the remaining terms that have already been included in the answer due to $\bar{C}D$.

12A

	1✓	4	5✓	9✓	13✓	14	15
*$\bar{C}D$	X		X	X	X		
$\bar{A}B\bar{C}$		X	X				
ABD					X		X
ABC						X	X

13Q $\bar{C}D$ is marked by * to indicate that it has been selected. Going from left to right the second necessary prime implicant is______.

13A $\bar{A}B\bar{C}$

This prime implicant is the only one with an X in the second column.

	1✓	4✓	5✓	9✓	13✓	14	15
*$\overline{C}D$	X		X	X	X		
*$\overline{A}B\overline{C}$		X	X				
ABD					X		X
ABC						X	X

14Q Note than an * is placed in front of $\overline{A}B\overline{C}$, and a check mark next to 4. The only other term which should be checked because of the selection of $\overline{A}B\overline{C}$ is standard term 5, but 5 has already been checked. Continuing to the right, the next necessary prime implicant is ______.

14A ABC

This prime implicant is the only one with an X under 14.

	1✓	4✓	5✓	9✓	13✓	14✓	15✓
*$\overline{C}D$	X		X	X	X		
*$\overline{A}B\overline{C}$		X	X				
ABD					X		X
*ABC						X	X

15Q Prime implicant ABC includes both 14 and 15; thus all standard sum terms have been checked. Since these three prime implicants include *all* standard sum terms no more prime implicants are necessary. Writing the final answer gives:

$X =$ ____________________

15A $X = \overline{C}D + \overline{A}B\overline{C} + ABC$

The only prime implicant not needed in the answer is ABD. It is redundant because 13 is covered by $\overline{C}D$, 15 is covered by ABC, and these are the only two X's in ABD.

16Q For this prime-implicant chart, $Y =$ ____________________

	5	7	9	12	13
$\bar{A}BD$	X	X			
$B\bar{C}D$	X				X
$A\bar{C}D$			X		X
$AB\bar{C}$				X	X

16A $Y = \bar{A}BD + A\bar{C}D + AB\bar{C}$

$\bar{A}BD$ is selected because of 7. $A\bar{C}D$ is selected because of 9. $AB\bar{C}$ is selected because of 12. These three cover the entire function (all columns).

	5✓	7✓	9✓	12✓	13✓
*$\bar{A}BC$	X	X			
$B\bar{C}D$	X				X
*$A\bar{C}D$			X		X
*$AB\bar{C}$				X	X

17Q

	6	7	10	11	14
$\bar{A}BC$	X	X			
$BC\bar{D}$	X				X
$A\bar{B}C$			X	X	
$AC\bar{D}$			X		X

The two prime implicants which *have* to be included in the answer are ______ and ______.

17A $\bar{A}BC$ (because of 7); $A\bar{B}C$ (because of 11)

18Q

	6 ✓	7 ✓	10 ✓	11 ✓	14
*$\bar{A}BC$	X	X			
$BC\bar{D}$	X				X
*$A\bar{B}C$			X	X	
$AC\bar{D}$			X		X

Column 14 has not yet been covered. How can it be covered?

18A By selecting *either* $BC\bar{D}$ or $AC\bar{D}$

19Q In this example, two prime implicants were required: $\bar{A}BC$ and $A\bar{B}C$. In addition, either $BC\bar{D}$ or $AC\bar{D}$ has to be selected. This gives rise to two different forms of Z:

19A $Z = \bar{A}BC + A\bar{B}C + BC\bar{D}$ and $Z = \bar{A}BC + A\bar{B}C + AC\bar{D}$

---

20 Summary

To select a set of prime implicants these steps are taken:

1. Search for columns with one X to find the necessary prime implicants.

2. Check off all terms included in the necessary prime implicants.

3. If a term of the standard sum has not been checked, several different solutions are possible.

Each solution contains a different prime implicant which includes the unchecked term.

21Q You are now equipped to work through a complete problem using both the Quine-McCluskey method and the prime-implicant chart. The problem is given in standard sum form and placed in the proper format. If it is not in this format you should be able to get it there. Write your answer in equation form. [*Hint:* There should be one term (repeated once) in the third list.]

$$X(A, B, C, D) = \Sigma(6, 7, 9, 11, 13, 15)$$

6	0110
9	1001
7	0111
11	1011
13	1101
15	1111

21A Given: $X(A, B, C, D) = \Sigma(6, 7, 9, 11, 13, 15)$

6	0110✓	6, 7	011-	9, 11, 13, 15	1--1
9	1001✓	9, 11	10-1✓	~~9, 11, 13, 15~~	~~1--1~~
7	0111✓	9, 13	1-01✓		
11	1011✓	7, 15	-111		
13	1101✓	11, 15	1-11✓		
15	1111✓	13, 15	11-1✓		

Prime implicants:

6, 7	$\bar{A}BC$
7, 15	BCD
9, 11, 13, 15	AD

Prime implicant chart:

	6✓	7✓	9✓	11✓	13✓	15✓
$*\bar{A}BC$	X	X				
BCD		X				X
$*AD$			X	X	X	X

Choose $\bar{A}BC$ because of 6, check 6 and 7. Choose AD because of 9, 11, 13, check 9, 11, 13, 15. BCD is redundant. The answer is $X = \bar{A}BC + AD$

22Q It has been pointed out that some problems have several answers. Following along with the premise that ultimately cost should be minimized, the final choice as to which solution to use depends on the cost of implementation.

$$X_1 = \bar{A}\bar{B} + \bar{A}C + AB\bar{C} + \bar{A}D$$
$$X_2 = \bar{A}\bar{B} + \bar{A}C + AB\bar{C} + B\bar{C}D$$

These two equations are two solutions obtained from a prime-implicant chart. Implement both functions and determine which one is cheaper in terms of diodes and transistors. Assume that all input variables are available in their *true forms only*.

22A X_1 is cheaper.

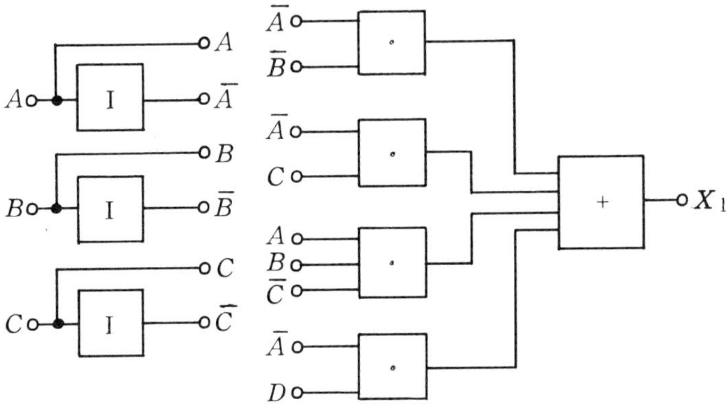

X_1 requires
3 transistors, 13 diodes

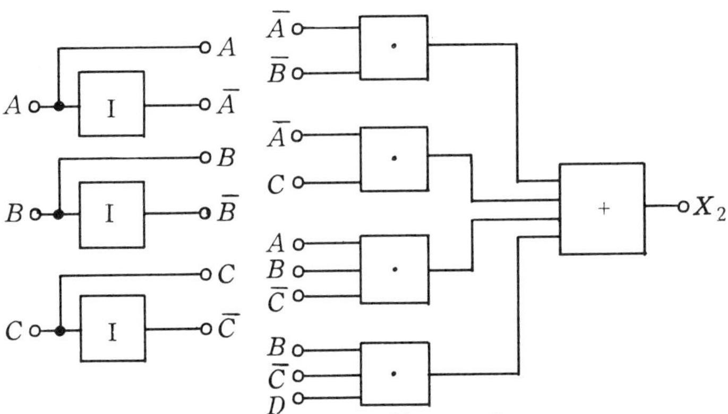

X_2 requires
3 transistors, 14 diodes

---

23 Summary of Quine-McCluskey method and the prime-implicant chart

1. Write the expression in ***standard sum*** form.
2. Translate to ***binary notation.***
3. List binary terms by *index*. (Sort to index groups, sort groups.)
4. Place *decimal equivalent* in front of each binary term.
5. *Compare* all pairs of terms whose indexes differ by 1 and *combine terms* which differ in one binary number, placing the new term with its decimal identification in the next list.
6. List *prime implicants* (unchecked terms).
7. Set up *prime implicant chart.*
   a. Terms of standard sum along the top.
   b. Prime implicants along the side.
   c. Place X's in chart.
8. Locate the *necessary* prime implicants by looking for columns with one X, checking the appropriate terms.
9. Find *all* combinations of the prime implicants which include the unchecked terms.
10. Compute the cost of the different solutions and choose the cheapest.

PROBLEMS

1. The following list represents the prime implicants of an expression. Find the minimum set of prime implicants from this list by use of the prime-implicant chart. Write your answer in equation form in terms of A, B, C, and D.

0	0000
5, 7	01-1
5, 13	-101
9, 13	1-01
10, 14	1-10
12, 13	110-
12, 14	11-0

2. Repeat Prob. 1 but with the following list of prime implicants. (In this problem you will have to make a choice early in the problem. This type of prime-implicant chart is referred to as cyclic.)

12, 13	110-
5, 13	-101
5, 7	01-1
6, 7	011-
6, 14	-110
12, 14	11-0

3. Minimize $X(A, B, C, D) = \Sigma(2, 3, 4, 5, 9, 10, 11, 13)$ and implement your answer.

4. Reduce the cost of the following circuit using the Quine-McCluskey method. (In this problem you will have to first place the expression in standard sum form. The method requires this.)

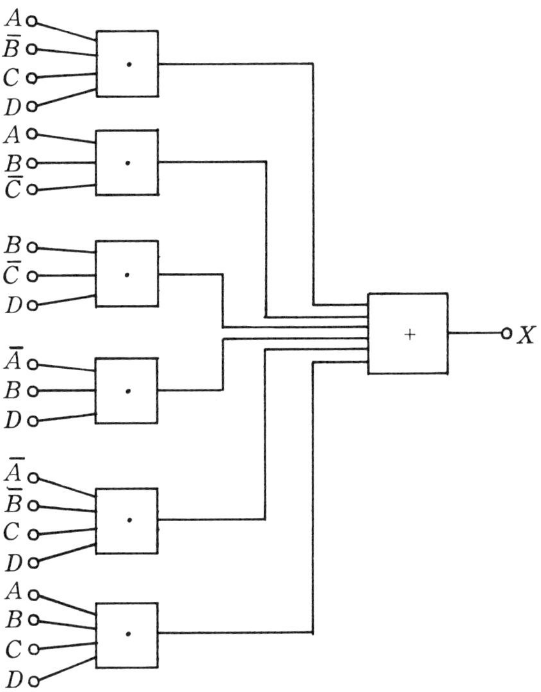

5. Design a circuit (output Z), with the aid of the Quine-McCluskey method, which compares the size of two binary numbers, each of which is two binary digits long. Assume that variables A and B represent one number, variables C and D the other number. The output of the circuit should be 1 if the first number (AB) is greater than or equal to the second number (CD). (*Hint:* Use a four-variable truth table to set up the problem. Plot the desired output on the table, read the table, and proceed from there.)

Chapter 10

KARNAUGH MAP—I

This chapter introduces the Karnaugh map method of graphically displaying a Boolean function. The construction of a map is shown, along with the procedure for plotting a function on the map. Chap. 11 will discuss how a Karnaugh map is used for minimization.

---

1 The Karnaugh map is a visual plot of Boolean functions. Its main purpose is to simplify these functions. Basically it operates much like the Quine-McCluskey method and the prime-implicant chart put together. First the map is used to determine the prime implicants of a function, and then it is used to determine which of these prime implicants are required and which are redundant.

The main disadvantage of the map is that it can only handle problems with at most five or six variables. For a larger number of variables, maps can be drawn and methods have been proposed, but the process necessary to obtain the answer is very involved. In this book most problems are limited to four variables except for a few examples in the problem section.

---

2 Each row in a truth table is devoted to one possible value a variable can take on. The one-variable table has two rows, one for each of the two possible values for that variable.

A	
0	
1	

Instead of allowing one row for each value of the variable, as in the table, the Karnaugh map allows one area for each value of the variable.

A	0	1

3Q The truth table is the plot of a function. It gives the value of a dependent variable as a function of the independent variables. The same is true of the map. For example, the truth table corresponds to the function on the map.

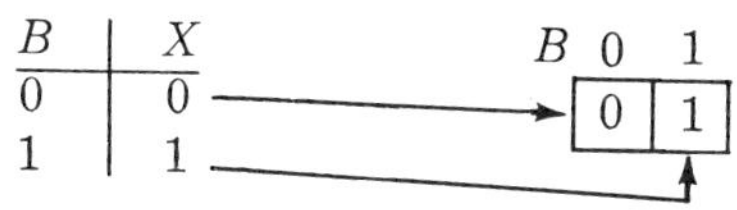

Table for *X* Map for *X*

Complete the map to correspond to the table:

A	*Y*
0	1
1	0

A	0	1

3A

A	0	1
	1	0

A 0 is placed in the square corresponding to A = 1. A 1 is placed in the square corresponding to A = 0. Superimposing corresponding maps gives the result.

A	0	1
		0

A	0	1
	1	

A	0	1
	1	0

4 The two-variable truth table has four rows because two variables can have four combinations of values. For the same reason the two-variable Karnaugh map has four squares. If a table and a map are written for the same variables, one row in the table corresponds to one square in the map. A two-variable Karnaugh map is shown below.

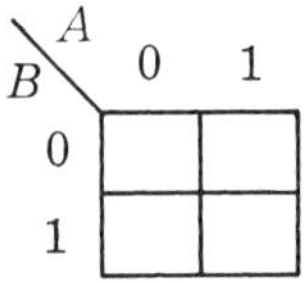

Any square in the table is identified by both the column and the row it is in. Since the top left square is in the $A = 0$ column and the $B = 0$ row, it will contain the first row in a truth table where the variables A and B are both 0.

5Q The top left square in the Karnaugh map corresponds to the condition $A = 0$, $B = 0$. The remaining squares correspond to:

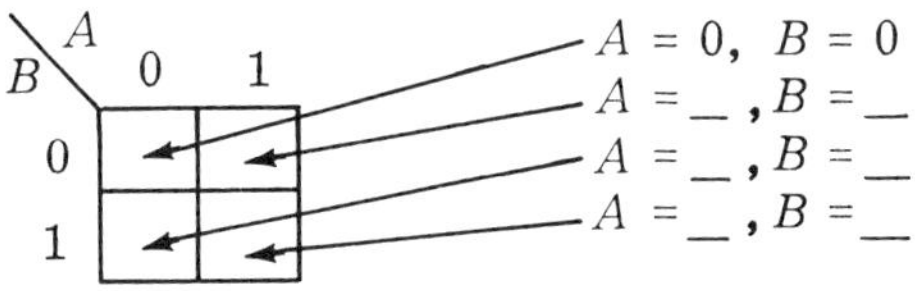

5A

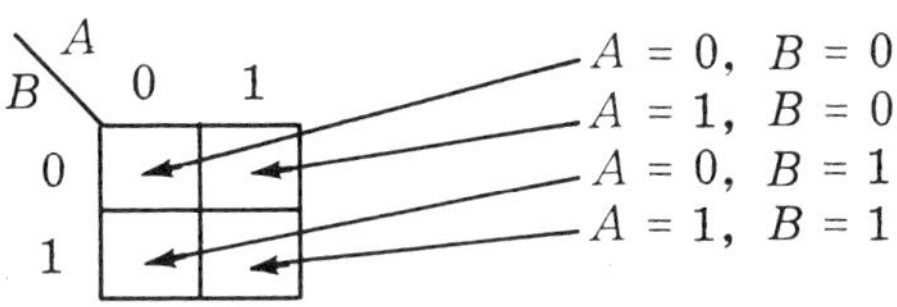

6Q Let us try to transpose a function from the table to the map. Fill in the squares of the map with the proper value.

A	B	Z
0	0	0
0	1	1
1	0	1
1	1	1

Table for Z

$B \backslash A$	0	1
0		
1		

Map for Z

$Z = A + B$

6A

$B \backslash A$	0	1
0	0	1
1	1	1

The top left square contains the value of Z if $A = 0$ and $B = 0$. According to the first row of the table, this value is 0, so a 0 is placed in the top left square. The bottom left square contains the value of Z when $A = 0$ and $B = 1$. This value is shown to be 1 in the second row of the table. The other entries are found in the same way.

7Q Transpose the function W from the table to the map.

$$W = A\overline{B} + \overline{A}B$$

A	B	W
0	0	0
0	1	1
1	0	1
1	1	0

$B \backslash A$	0	1
0		
1		

7A

$B \backslash A$	0	1
0	0	1
1	1	0

The two rows of the table for which the value of $W = 1$ are $A = 0$, $B = 1$ and $A = 1$, $B = 0$. These two rows correspond to the bottom left and top right square on the map. The remaining squares should contain 0's.

8 The three-variable map contains eight squares just as the three-variable table contains eight rows.

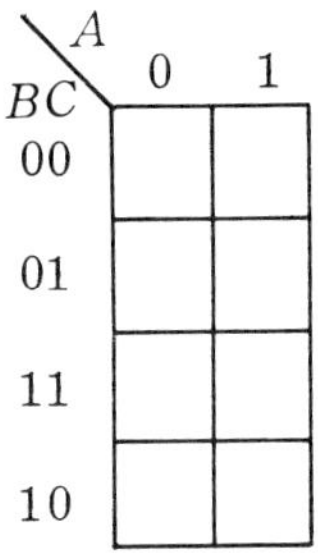

Each row in the map now represents one of the four possible combinations of values of the two variables B and C. The two columns still represent the two states of the variable A.

It should be pointed out that the order of the rows in the map does not correspond to the order in which the rows appear on the table. This peculiar order of rows is chosen because when moving from one row to its adjacent one only one variable should change its value. For example, going from 11 to 01 only B changes. This point is only brought to your attention now. In the second half of this chapter and in Chap. 11 we will make use of this property.

9Q As an example, let us locate the square corresponding to $A = 1$, $B = 0$, $C = 1$. It has to be in the $A = 1$ column (shaded column below). It also has to be in the row where $B = 0$ and $C = 1$, which is indicated as the 01 row (shaded row). The cross hatched area corresponds to $A = 1$, $B = 0$, $C = 1$.

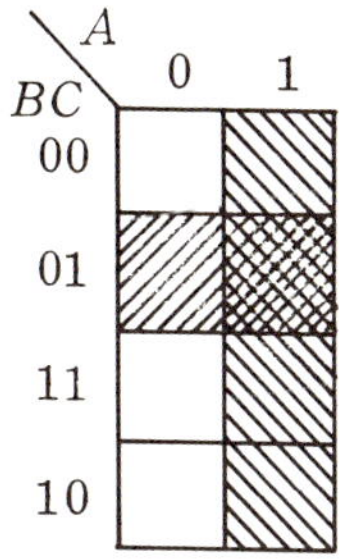

Mark the square on the map which corresponds to $A = 0$, $B = 1$, $C = 1$.

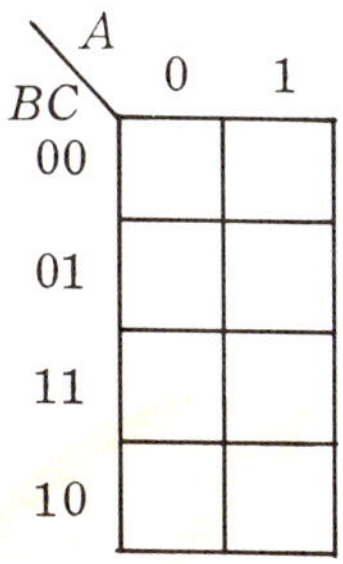

9A

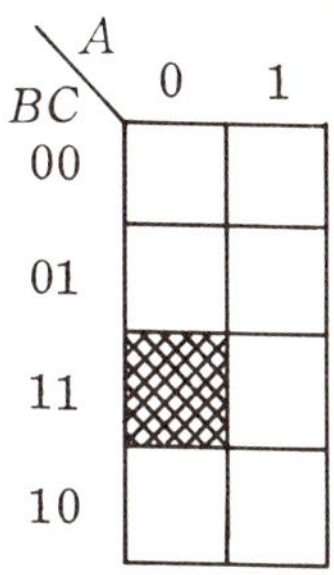

$A = 0$ specifies the left column; $B = 1$ and $C = 1$ specify the third row.

10Q Mark the square for which $A = 1$, $B = 1$, $C = 0$.

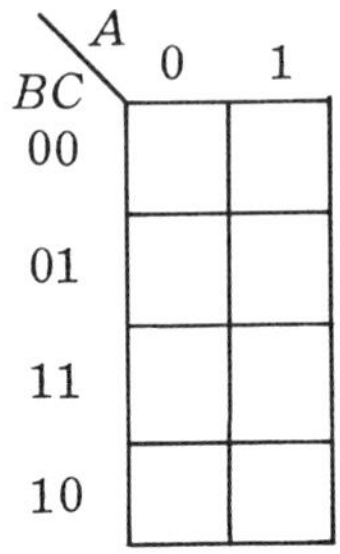

10A

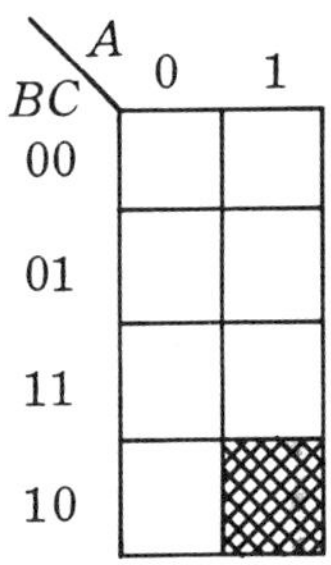

11Q X is plotted on the table below; transpose it to the map.

A	B	C	X
0	0	0	0
0	0	1	1
0	1	0	1
0	1	1	1
1	0	0	0
1	0	1	1
1	1	0	0
1	1	1	0

Table for X

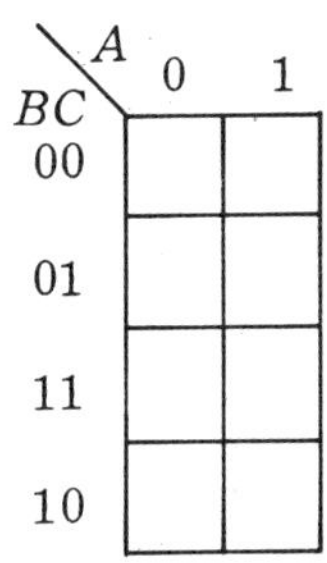

Map for X

11A

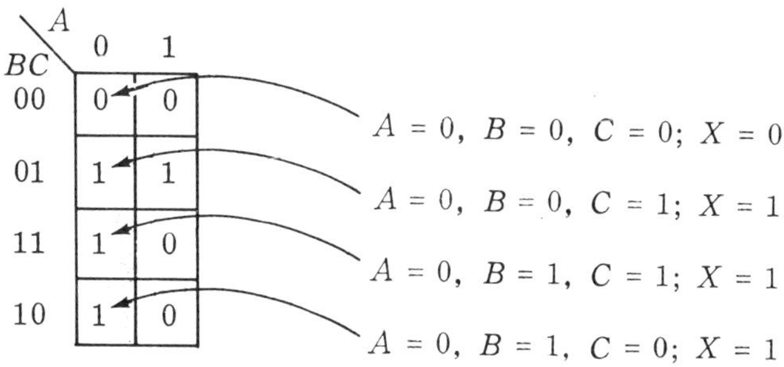

12 A four-variable map has sixteen squares arranged in four rows and four columns. Note that the order of the columns is the same as the order of the rows: 00, 01, 11, 10.

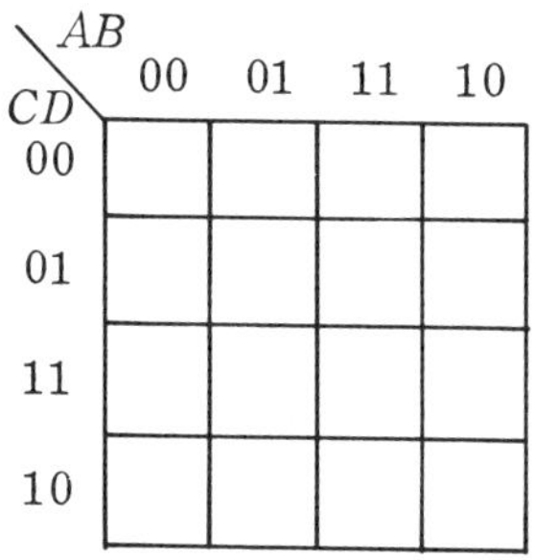

13Q Locating a square on the four-variable map is similar to locating a square on a map of a different size. For example, locate the square for:

$A = 0$
$B = 1$
$C = 1$
$D = 1$

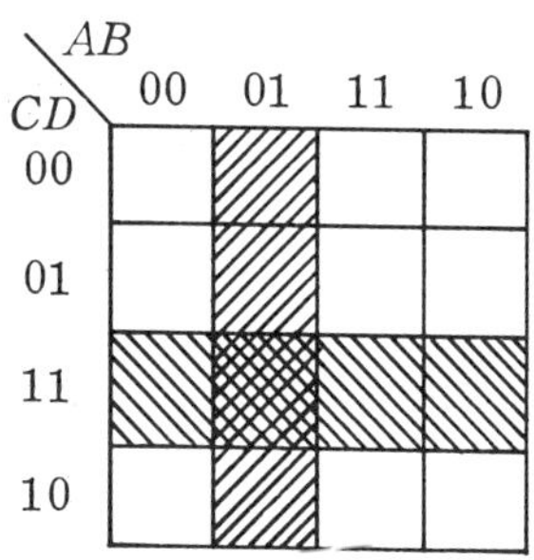

It must be in the column where $A = 0$ and $B = 1$. That column is 01. It must also be in the row where $C = 1$ and $D = 1$. That row is 11. The intersection of the two is the desired square. Locate the square for:

$A = 0$
$B = 1$
$C = 0$
$D = 1$

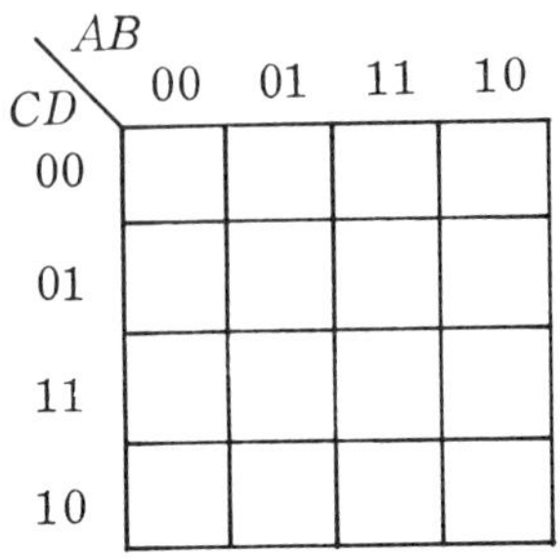

13A

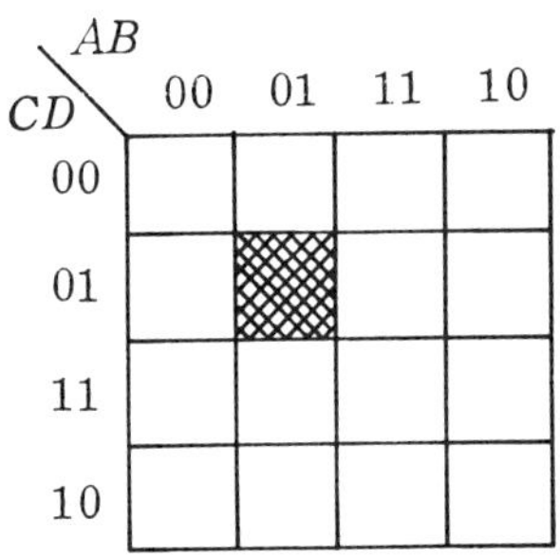

---

14 Summary

Thus far the one-, two-, three-, and four-variable Karnaugh maps have been introduced. They are repeated here.

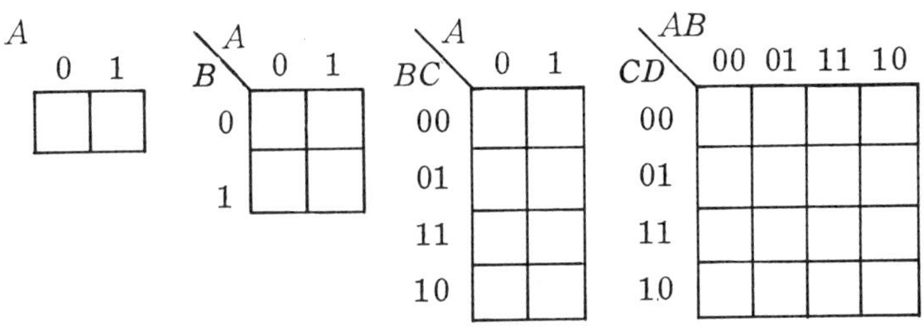

To draw a five- (or six-) variable map, two (or four) four-variable Karnaugh maps are used.

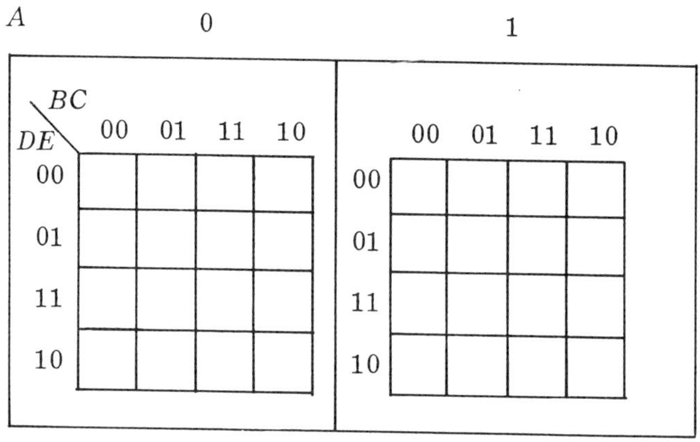

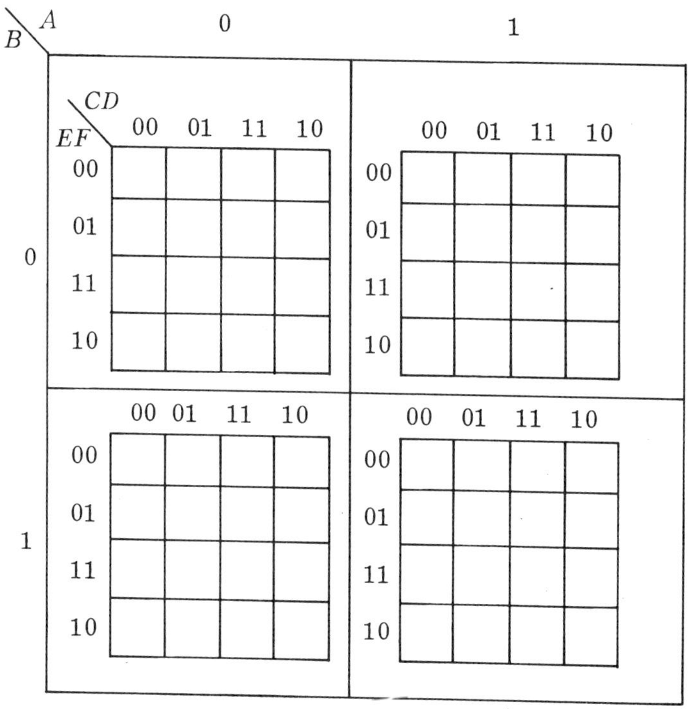

Many other shapes of these maps have been proposed and are used. These two shapes appear to be the most popular. Basically the five- (six-) variable map is a one- (two-) variable map where each square contains a four-variable map.

15Q Up until now, we have filled in maps only from a truth table. In the remainder of this chaper we will plot functions directly on the maps. Choose and draw the proper size map and plot the function X; plot where X is a function of the variables A, B, C, and D.

$$X = A\overline{B}\overline{C}D + ABC\overline{D}$$

15A

$CD \backslash AB$	00	01	11	10	
00	0	0	0	0	
01	0	0	0	1	← $A = 1,\ B = 0,\ C = 0,\ D = 1$
11	0	0	0	0	
10	0	0	1	0	

The first term $A\overline{B}\overline{C}D$ states that X is 1 if $A = 1$, $B = 0$, $C = 0$, and $D = 1$; thus a 1 is entered in the map at that square. The second term $ABC\overline{D}$ states that $X = 1$ if $A = 1$, $B = 1$, $C = 1$, and $D = 0$, so a 1 is placed in that square also. Since the function is 0 for all other conditions, a 0 is placed in all other squares. A four-variable map is used because X is a function of four variables, namely, A, B, C, and D.

---

16 A function is completely specified if each square in a map is specified. To save time, however, we will fill in only the 1's of a map and assume that all empty squares are actually 0's. (*Note:* Maps should only contain 0's, 1's, and, later, loops. No check marks or other characters should be entered.)

CD \ AB	00	01	11	10
00	0	1	0	0
01	0	0	1	0
11	0	1	0	1
10	0	0	0	0

=

CD \ AB	00	01	11	10
00		1		
01			1	
11		1		1
10				

17 Thus far functions in standard sum form were plotted on the map. The following example will describe the plotting of nonstandard form functions. The functional notation will be used again. $Y(A, B, C, D)$ means that Y is a function of the four variables A, B, C, and D. (It was first introduced in Chap. 4, frame 21, page 59.

Example

$$Y(A, B, C, D) = \bar{A}\bar{B}C$$

A 1 should be placed in all squares where $A = 0$, $B = 0$, and $C = 1$. $A = 0$ and $B = 0$ together specify the first column, and $C = 1$ in the third and fourth rows. The intersection of the areas covers the squares for which A and B are 0 and C is 1.

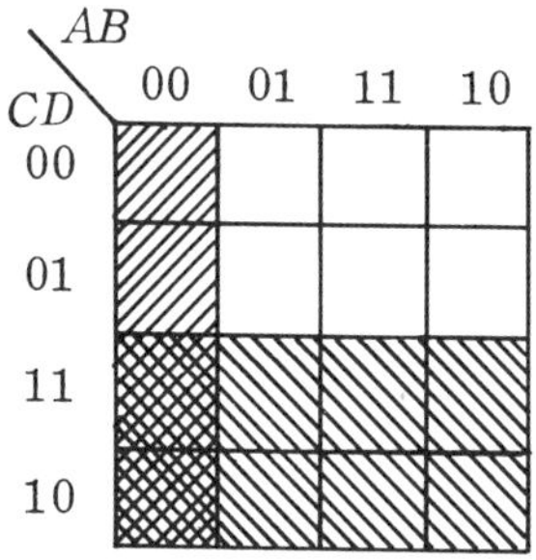

CD \ AB	00	01	11	10
00				
01				
11	1			
10	1			

18Q Plot $Z(A, B, C, D) = AB\bar{C}$ on the proper size map. [*Hint:* To locate squares, first find the correct column(s) and then the correct row(s).] Braces are often used along the sides to help the location of squares.

18A

AB (column 11)

$CD \backslash AB$	00	01	11	10
00			1	
01			1	
11				
10				

$\bar{C}$ (rows 00, 01)

19Q Plot $X = \bar{A}C\bar{D}$ on the four-variable map.

$CD \backslash AB$	00	01	11	10
00				
01				
11				
10				

19A

$CD \backslash AB$	00	01	11	10
00				
01				
11				
$C\bar{D}$ 10	1	1		

$\bar{A}$ (columns 00, 01)

20 Functions thus far have had only one term. When a function contains more than one term, the occasion of the 1's is determined separately for each term, and the final map contains the 1's from all the terms. For example,

Plot $X(A, B, C, D) = AB + \bar{A}\bar{B}C$

To demonstrate which 1's come from which terms, the two terms are first plotted on separate maps, and then the final map is shown with all the 1's from both terms.

CD \ AB	00	01	11	10
00			1	
01			1	
11			1	
10			1	

Map for AB

CD \ AB	00	01	11	10
00				
01				
11	1			
10	1			

Map for $\bar{A}\bar{B}C$

CD \ AB	00	01	11	10
00			1	
01			1	
11	1		1	
10	1		1	

Map for $AB + \bar{A}\bar{B}C$

21Q Plot the function Z on the map shown below:

$$Z = A\overline{B}CD + \overline{A}BC$$

CD \ AB	00	01	11	10
00				
01				
11				
10				

21A

CD \ AB	00	01	11	10
00				
01				
11		1		1
10		1		

22Q Plot Y on the proper size map.

$$Y(A, B, C) = AB + \overline{B}\overline{C}$$

22A

BC \ A	0	1
00	1	1
01		
11		1
10		1

A three-variable map had to be used because Y is a function of three variables.

23 Sometimes two terms generate 1's in the same square. In this case, a single 1 is placed in the square. This is essentially the same as performing the operation $1 + 1 = 1$. For example,

$$X(A, B, C) = AC + BC$$

BC \ A	0	1
00		
01		1
11		1
10		

Map of AC

BC \ A	0	1
00		
01		
11	1	1
10		

Map of BC

BC \ A	0	1
00		
01		1
11	1	1
10		

Map of $AC + BC$

24Q Plot the function Y:

$$Y = AB + CD$$

CD \ AB	00	01	11	10
00				
01				
11				
10				

24A

CD \ AB	00	01	11	10
00			1	
01			1	
11	1	1	1	1
10			1	

CD \ AB	00	01	11	10
00			1	
01			1	
11			1	
10			1	

Map for AB

CD \ AB	00	01	11	10
00				
01				
11	1	1	1	1
10				

Map for CD

Superimposing the two separate maps for AB and CD gives the answer.

25Q You have had considerable practice with functions containing one and two terms. Now more complex functions will be plotted. Take your time on these problems; take one term at a time.

$$X = \bar{A}BC + A\bar{B}C + AB\bar{C} + BCD$$

CD \ AB	00	01	11	10
00				
01				
11				
10				

25A

$CD \backslash AB$	00	01	11	10
00			1	
01			1	
11		1	1	1
10		1		1

Map for X

$CD \backslash AB$	00	01	11	10
00				
01				
11		1		
10		1		

$\bar{A}BC$

$CD \backslash AB$	00	01	11	10
00				
01				
11				1
10				1

$A\bar{B}C$

$CD \backslash AB$	00	01	11	10
00			1	
01			1	
11				
10				

$AB\bar{C}$

$CD \backslash AB$	00	01	11	10
00				
01				
11		1	1	
10				

BCD

26Q Plot the function $Z(A, B, C, D) = \overline{B}$. (*Hint:* This function has eight 1's.)

$CD \backslash AB$	00	01	11	10
00				
01				
11				
10				

26A

$CD \backslash AB$	00	01	11	10
00	1			1
01	1			1
11	1			1
10	1			1

27 It is a property of the map that $\bar{B}$ appears in the left-most column and the right-most column. $\bar{D}$ appears in the top and bottom rows.

CD \ AB	00	01	11	10
00				
01				
11	1			1
10				

$\bar{B}CD$

CD \ AB	00	01	11	10
00	1			
01				
11				
10	1			

$\bar{A}\bar{B}\bar{D}$

CD \ AB	00	01	11	10
00	1			1
01				
11				
10	1			1

$\bar{B}\bar{D}$

The last term is by far the most difficult one. It is the one missed by many students.

28Q Plot $U(A, B, C, D) = AB\overline{D} + \overline{B}CD + \overline{A}\,\overline{B}D$

CD \ AB	00	01	11	10
00				
01				
11				
10				

28A

CD \ AB	00	01	11	10
00			1	
01	1			
11	1			1
10			1	

Map for U

CD \ AB	00	01	11	10
00			1	
01				
11				
10			1	

$AB\overline{D}$

CD \ AB	00	01	11	10
00				
01				
11	1			1
10				

$\overline{B}CD$

CD \ AB	00	01	11	10
00				
01	1			
11	1			
10				

$\overline{A}\,\overline{B}D$

29 Summary

Functions were plotted on a Karnaugh map by plotting one term at a time. Each term is found by:

1. Determining the column(s) according to the column headings
2. Determining the row(s) according to the row headings
3. Placing 1's in all squares covered by the intersection of the selected row(s) and column(s)

An interesting fact is that when plotting individual terms on the map, the number of 1's described by a term is always a power of 2. If M is the number of variables for which the map is drawn, N the number of variables in the term, and O the number of 1's described by the term, then the following relation holds:

$$O = 2^{(M-N)}$$

For example, for a four-variable map ($M = 4$)

N	O
1	8
2	4
3	2
4	1

PROBLEMS

1. Plot the following functions on Karnaugh maps of the proper size. You may have to change certain functions to the proper form.
   (a) $U(A, B, C, D) = A\overline{B} + A\overline{C} + BD$
   (b) $V(A, B, C) = \Sigma(000, 010, 111)$
   (c) $W(A, B, C, D) = \Sigma(0, 2, 4, 5, 9, 12, 15)$
   (d) $X = (A + \overline{B})C + (B + \overline{D})A$
   (e) $Y = (A + B)(\overline{A} + \overline{B})$
   (f) $\overline{Z} = (\overline{A} + C + \overline{D})(A + \overline{B} + C)(\overline{C} + D)$
   (Plot Z, *do not* plot $\overline{Z}$.)

2. Design a gate circuit which has two data inputs (D_1, D_2), two control inputs (C_1, C_2), and one output (X).

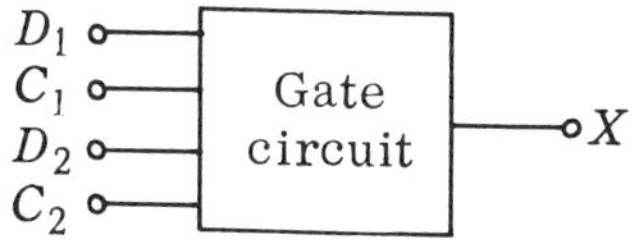

If both control inputs are 0, the output should be 0. If $C_1 = 1$ and $C_2 = 0$ the output should equal D_1, and if $C_2 = 1$ and $C_1 = 0$ the output should equal D_2. If both C_1 and C_2 are 1 at the same time, the output is the inverse of D_1.

(*a*) Plot function X on a truth table (use the following order of variables: D_1, C_1, D_2, C_2).
(*b*) Write an equation for X.
(*c*) Simplify the equation for X.
(*d*) Implement the expression for X.
(*e*) Plot on the Karnaugh map using the table from Prob. 2*a*.

3. A TRUE-INVERT gate has three inputs and one output Z. One input is the data line D, the second is the inverted data $\overline{D}$, and the third is a control line C. Z is equal to D if $C = 1$, Z is equal to $\overline{D}$ if $C = 0$. Plot Z as a function of C and D on the appropriate size map. (*Hint:* If you have trouble, work Prob. 1*a* and *e* first.

4. A four-input, one-output circuit is to be designed. Three of the four inputs (variables A, B, and C) are the outputs of a binary counter. (The variables represent a binary number.) Assume that variable A represents the high-order digit of the number, variable C the low-order digit of the number. The output for the circuit should be 1 if the fourth input variable (D) is 1 and the binary number at the input (represented by A, B, and C) is a 0, 5, or 7. Obtain the values for the output (X) on a map. (*Hint:* If $A = 0$, $B = 1$, $C = 1$, the binary number represented by A, B, and C is 011, which is 3.)

5. Instructions are stored in a computer as binary numbers. For example, the instruction ADD may be stored as 011, which can be broken down as $A = 0$, $B = 1$, $C = 1$; the instruction SUBTRACT as 100, $A = 1$, $B = 0$, $C = 0$; and the instruction MULTIPLY as 101. Plot a function X on a Karnaugh map which is 1 when a fourth input $D = 1$, *and* any one of the three instructions mentioned above is present. $X = 0$ for all other cases.

6. A five-variable map consists of two four-variable maps. In locating a square, one first decides which map to use, then proceeds within that chosen map. (Left map, $A = 0$; right map, $A = 1$.)

$A = 0$

DE \ BC	00	01	11	10
00			1	
01				
11				
10				

$A = 1$

DE \ BC	00	01	11	10
00				
01				
11				
10				

Locate the following squares (the first term is a worked example):

$\bar{A}BC\bar{D}\bar{E}$
$\bar{A}\bar{B}CDE$
$\bar{A}B\bar{C}DE$
$A\bar{B}C\bar{D}E$
$ABCDE$

Chapter 11

KARNAUGH MAP—II

In Chap. 10 functions were plotted on two-, three-, and four-variable Karnaugh maps. It is the purpose of this chapter to develop a method of reading the map in such a way that a minimized expression is obtained. The expression consists of a minimum set of prime implicants; thus, this method gives the same solution as the Quine-McCluskey prime-implicant chart method does. In essence the same theorem is used in both methods except that on the map a visual picture is used instead of the algebra. The map is a simple and quick method, but it is limited to problems with less than five or six variables.

1Q The basic properties of the Karnaugh map are the following:

1. Each square on the map corresponds to one specific assignment of values to each variable.
2. For adjacent squares the assignment of only one variable is different.
3. Adjacent squares have one common side. (This point will be extended in frame 7.)

$CD \backslash AB$	00	01	11	10
00				
01				1
11				1
10				

The two squares marked with 1's are adjacent. The assignment of one variable is different. Which one is it?

1A C

2Q In each map two squares are marked. Are these 1's adjacent? If no, why not?

$CD \backslash AB$	00	01	11	10
00				
01		1		
11			1	
10				

Map 1

$CD \backslash AB$	00	01	11	10
00				
01	1			
11	1			
10				

Map 2

$CD \backslash AB$	00	01	11	10
00				
01		1	1	
11				
10				

Map 3

In map 1:______; in map 2:______; in map 3:______.

2A (1) No—assignment of two variables is different (A and C); (2) yes; (3) yes.

3Q

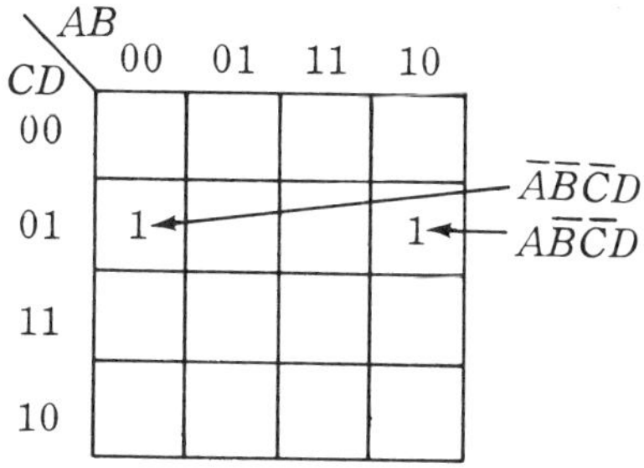

The two squares marked above (do) (do not) differ in the assignment of only one variable.

3A *do* differ in the assignment of only one variable

$\bar{A}\bar{B}\bar{C}D$ and $A\bar{B}\bar{C}D$: variable A is different

---

4 In these two maps adjacent squares are marked with arrows. You should note that squares on the ends of each row and each column *are adjacent.*

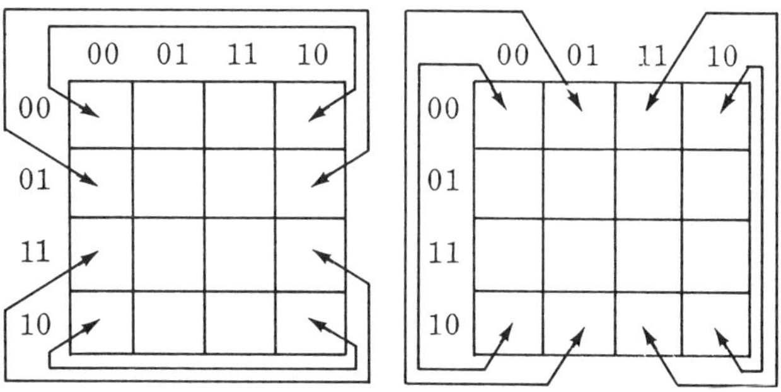

---

5 The definition for adjacency holds not only for the four-variable Karnaugh map, but for the two- and three-variable maps as well. This point will not be proved, but for example, in the three-variable map, $\bar{A}\bar{B}C$ and $\bar{A}BC$ are adjacent. In this map, again the adjacency is due to the sequence of row headings which are the same in the different size maps.

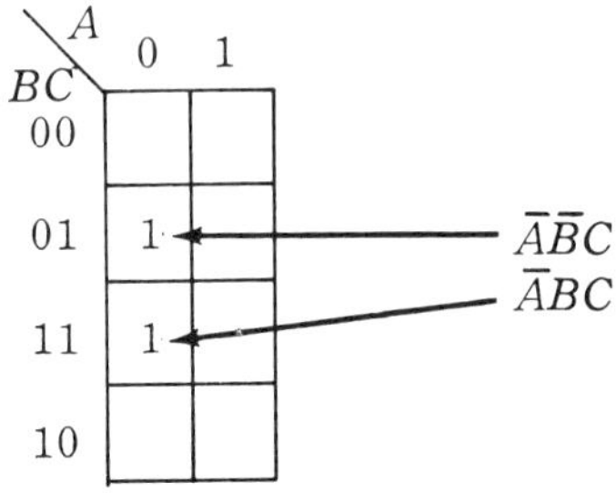

6 We now know two facts which we can put together.

1. The value of one variable is different for adjacent squares.
2. If two terms differ in the value of one variable, they can be combined to one term by using the theorem $AB + A\bar{B} = A$, where A can stand for several variables ANDed together. (Same as Quine-McCluskey method.)

Here is the general rule which can be used to read two adjacent squares.

Rule: Two adjacent squares can be read as one term by including only those variables whose value is the same for both squares.

7 **Example**

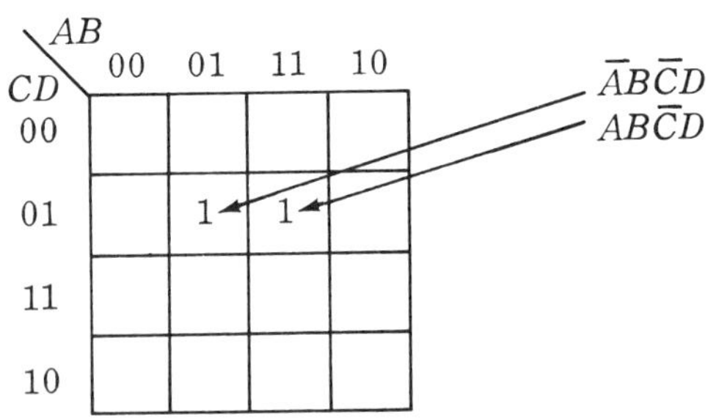

Applying the rule to this example we find that:

1. Value of A is different for the two squares.
2. B is in the true form for both squares.
3. C is in the inverted form for both squares.
4. D is in the true form for both squares.

Thus the term plotted on the map is $B\overline{C}D$. A was eliminated because its value is different for the two squares.

Compare this rule to Quine-McCluskey.

$\overline{A}B\overline{C}D$ corresponds to 0101
$AB\overline{C}D$ corresponds to 1101
The two numbers combine to -101
Written out in variables the new term corresponds to $B\overline{C}D$
Same result as from map.

8Q Function X is plotted on this map; read it.

CD \ AB	00	01	11	10
00	1			
01	1			
11				
10				

$X =$ ____________

8A $X = \overline{A}\,\overline{B}\,\overline{C}$

The value for D is different for the two squares; the values for $\overline{A}$, $\overline{B}$, and $\overline{C}$ are the same.

9 When reading functions from a map it is helpful to loop the 1's that were combined. As you will see later, loops may be of different sizes. If a loop contains two 1's it is called a 2-loop.

10Q Applying the process to the three-variable map, read Y. Note the looping of the 1's.

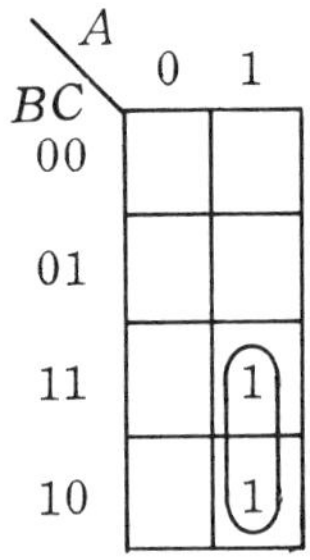

$Y =$ __________________

10A $Y = AB$

11Q Note that in the last frame the problem had three variables and the answer had only two variables. This is because a 2-loop on any size map has one less variable than the map.

If more than one pair of squares are plotted on one map they are read independently.

Example

In the left map on page 199 there are two pairs of 1's. Thus the function plotted on the map is $Z = A\overline{C}D + \overline{A}C\overline{D}$. The function X on the right map is equal to $X =$ __________.

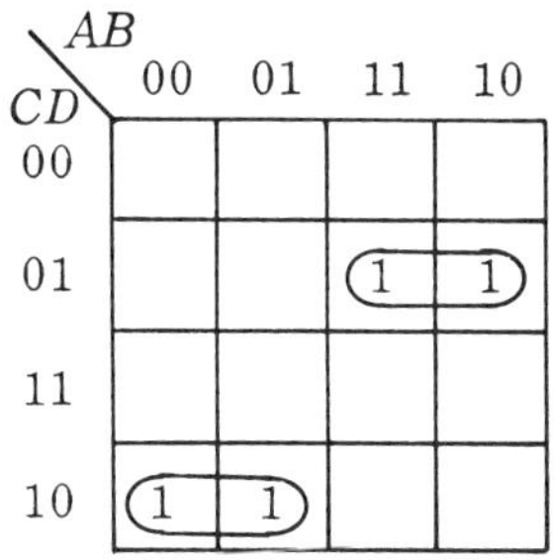

Map for Z

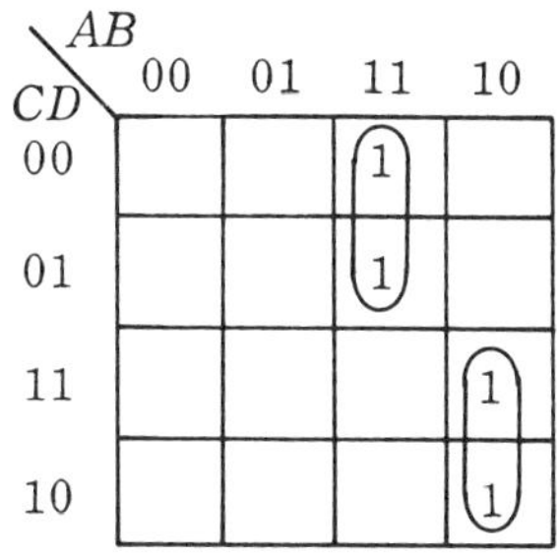

Map for X

11A $X = AB\overline{C} + A\overline{B}C$

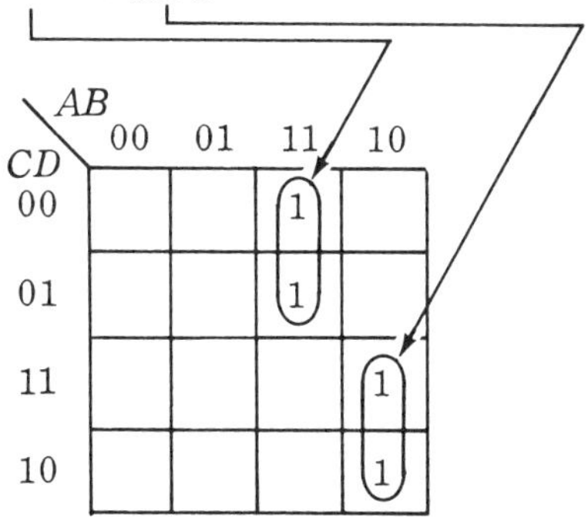

12Q Loop the map, remembering that only adjacent squares can be combined; squares touching on corners are *not* adjacent.

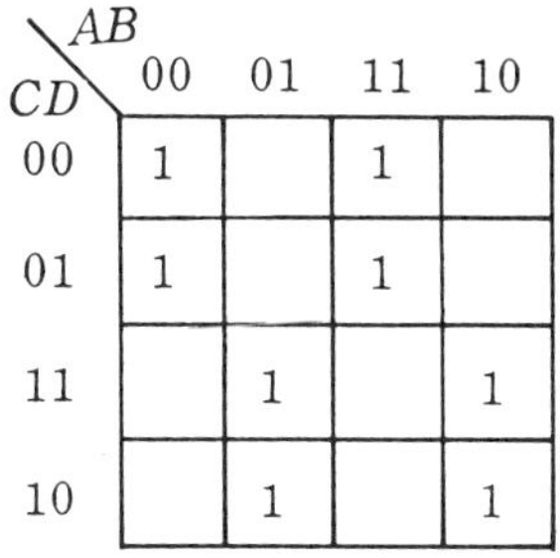

$X =$ ____________________

12A $X = \overline{A}\overline{B}\overline{C} + \overline{A}BC + AB\overline{C} + A\overline{B}C$

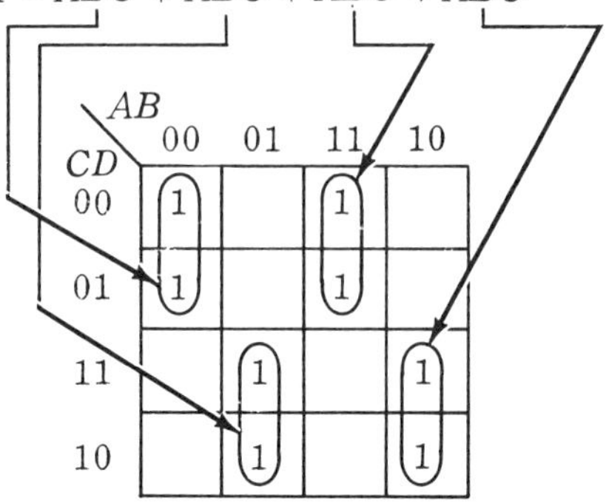

13

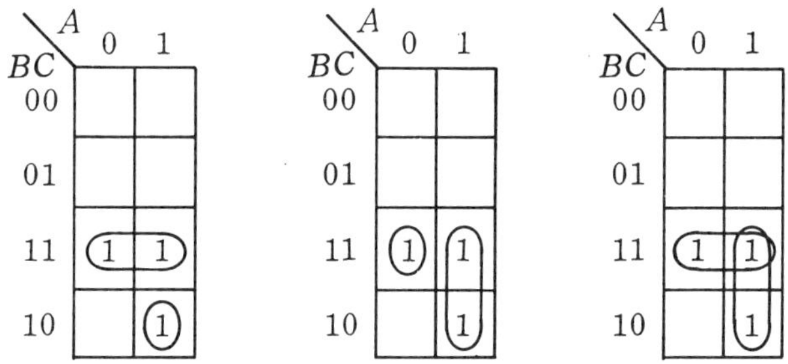

$Y = BC + AB\overline{C}$ $\quad$ $Y = \overline{A}BC + AB$ $\quad$ $Y = AB + BC$

The simplest form of Y is the one on the extreme right. In this map a 1 was used in two different loops. The following algebraic reasoning will demonstrate the validity of using a 1 more than once. Y in standard sum form is

$$Y = \overline{A}BC + AB\overline{C} + ABC$$

In any expression a term can be repeated as often as necessary, because of the theorem $A + A = A$. Looking at the equation for Y, we see that the last term can be combined with either of the other terms. To combine the expression into two simpler terms the last term is repeated.

$$Y = \bar{A}BC + AB\bar{C} + ABC + ABC$$

Combining terms:

$$Y = BC + AB$$

Repeating terms is the same as using a 1 in the map twice. The square ABC is used twice to give the two 2-loops.

If this problem would be worked in the Quine-McCluskey method the answer would be the same as shown here.

$$Y = \bar{A}BC + AB\bar{C} + ABC$$
$$Y(A, B, C) = 011 + 110 + 111$$

3	011✓	3, 7	-11
6	110✓	6, 7	11-
7	111✓		

$$Y = BC + AB$$

14Q *Rule:* A 1 on the map can be used in as many loops as needed.

It will be helpful in working the problems from this point to loop the 1's which have been read. It should be pointed out that the order in which terms are read on the map is of no significance because

$$A + B + C = A + C + B = B + A + C$$

Using this rule find the function plotted on the map below. *Hint:* Three 2-loops.

CD \ AB	00	01	11	10
00			1	
01		1	1	
11			1	
10				

U = ____________________

14A $U = B\bar{C}D + AB\bar{C} + ABD$

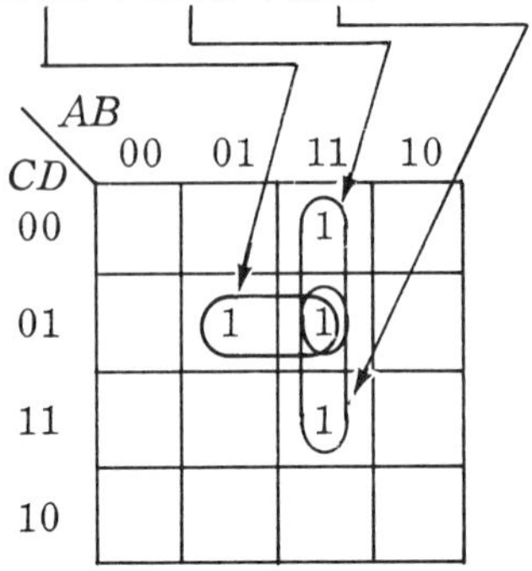

Did you loop the map properly?

15Q Read this function from the map. Mark loops on the map.

$V =$ ____________________

CD \ AB	00	01	11	10
00		1	1	
01	1			1
11				
10		1		

15A $V = \bar{B}\bar{C}D + \bar{A}B\bar{D} + B\bar{C}\bar{D}$

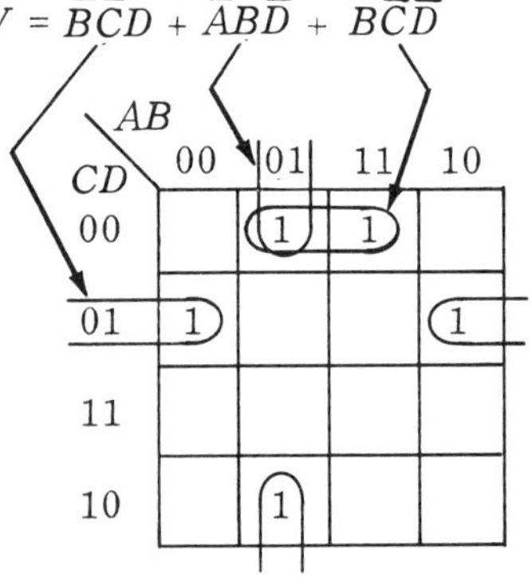

Here the square $\bar{A}B\bar{C}\bar{D}$ is used twice. Note the 2-loops which go from one edge to the other.

16Q In the last frame two loops were found which went around the back. These loops are usually indicated on the map by two U's open to the side of the map. (See frame 15A above.) Find *two horizontal 2-loops* which cover the map.

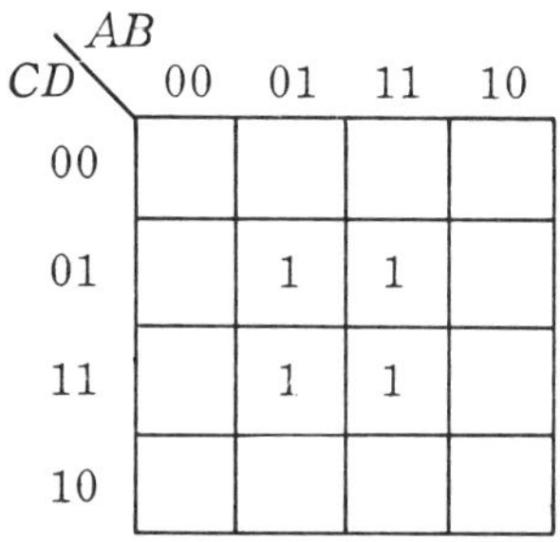

$X =$ ______________

16A $X = B\bar{C}D + BCD$

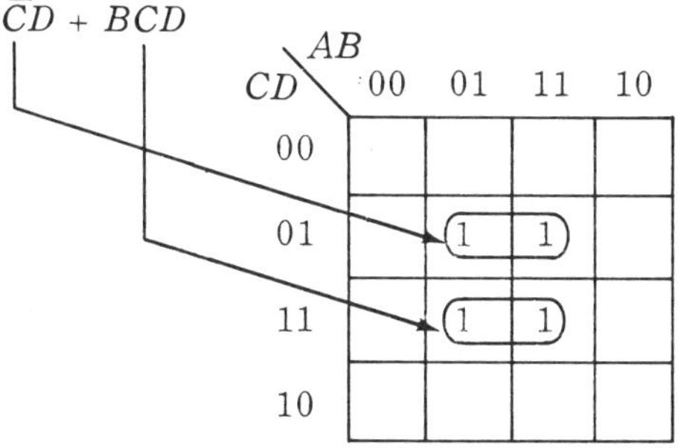

---

17 Another answer for X is:

$X = \bar{A}BD + ABD$

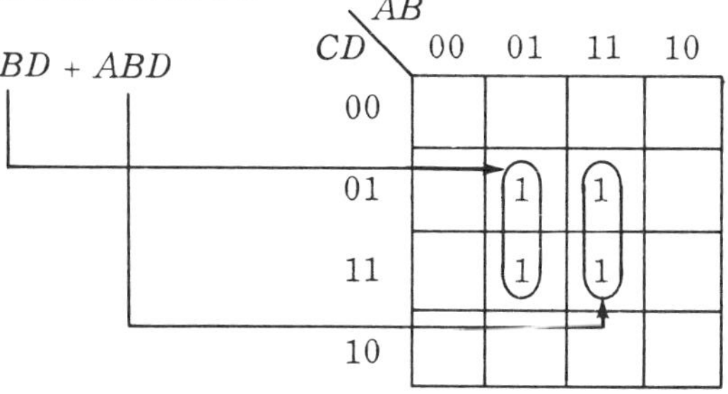

Note that there are two distinctly different equations describing this map.

$$X = B\bar{C}D + BCD$$
$$X = \bar{A}BD + ABD$$

Looking at these terms closely we find that in each equation the terms can be combined to BD. It can be concluded that these four 1's represent a larger loop, namely, the 4-loop BD.

A 4-loop is described by two variables less than the map has variables. For this example the map has four variables; therefore a 4-loop is described by $(4 - 2 = 2)$ two variables. When reading the 4-loop the same rule applies as for the 2-loops, namely:

Rule: The loop is described by the variables whose value remains constant for the entire loop.

This rule can be turned around to read that if four 1's *do not* have two variables whose value remains the same for *all* 1's, they do not represent a 4-loop.

---

18 Compare the result from frame 17 with the already familiar Quine-McCluskey method. The function on the map is

$$X = \bar{A}B\bar{C}D + \bar{A}BCD + AB\bar{C}D + ABCD$$

Karnaugh map

The map was read directly in 2-loops. Two answers were found:

$$X = B\bar{C}D + BCD$$
$$X = \bar{A}BD + ABD$$

or reading the 4-loop:

$$X = BD$$

Quine-McCluskey

First list:

5	0101
7	0111
13	1101
15	1111

Second list corresponds to 2-loops as shown in frame 7:

5, 7	01-1
5, 13	-101
7, 15	-111
13, 15	11-1

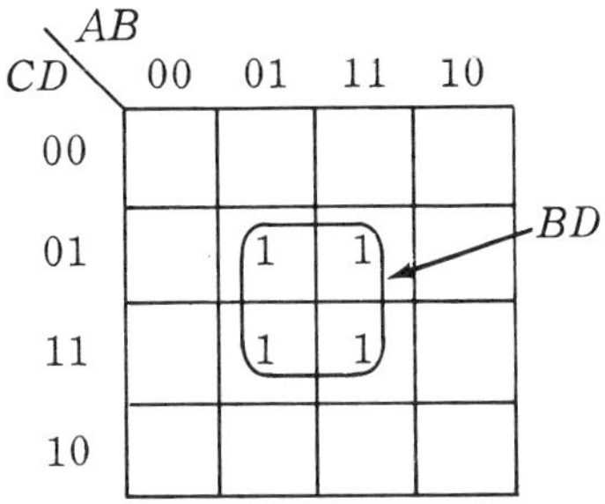

Note that if the prime-implicant chart would be applied now, two pairs of terms would be selected (5, 7 01-1 and 13, 15 11-1 or 5, 13 -101 and 7, 15 -111). Compare these two answers with the one obtained by the map. They are the same. Another list can be formed by combining terms in the second list.

5, 7, 13, 15	-1-1
5, 7, 13, 15	-1-1

In this list entries appear twice. The reason is now more easily seen. There are two different pairs of terms (like 2-loops) which combine to make one term in the third list. The third list corresponds to 4-loops.

19Q 4-loops are described by two less variables than the problem contains. The function on this map is:

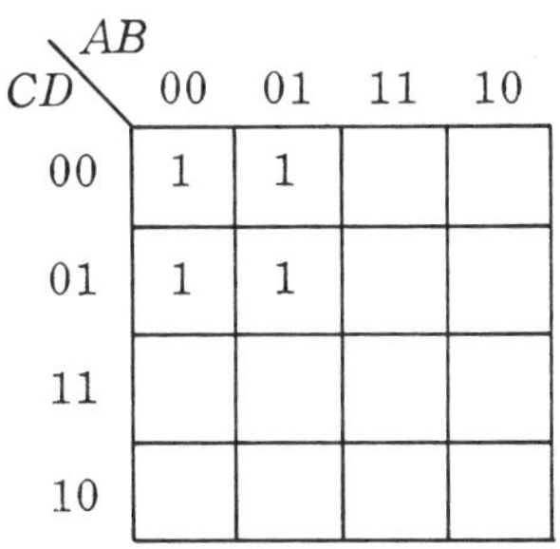

$Y =$ ________________

19A $Y = \bar{A}\bar{C}$

Value for A remains constant (0). Value for C remains constant (0). Four 1's with two variables constant throughout loop is 4-loop. Loop is $\bar{A}\bar{C}$.

20Q Which two variables remain constant in this 4-loop?

CD \ AB	00	01	11	10
00			1	
01			1	
11			1	
10			1	

$Z =$ ______________

20A $Z = AB$

Both A and B are constant; C and D change.

21 Basically the 4-loops can take on two different shapes. Here are some examples for both shapes:

1. Four in a square arrangement.

CD \ AB	00	01	11	10
00		1	1	
01		1	1	
11				
10				

Some place in the map

CD \ AB	00	01	11	10
00	1			1
01	1			1
11				
10				

Split and appears in both sides

CD \ AB	00	01	11	10
00		1	1	
01				
11				
10		1	1	

Split and appears on top and bottom

CD \ AB	00	01	11	10
00	1			1
01				
11				
10	1			1

The four corners; split both ways

2. Four in one column or one row.

CD \ AB	00	01	11	10
00			1	
01			1	
11			1	
10			1	

CD \ AB	00	01	11	10
00				
01	1	1	1	1
11				
10				

22Q On this map there is more than one 4-loop. Loop map as you read it.

CD \ AB	00	01	11	10
00		1		1
01		1		1
11		1		1
10		1		1

$Z =$ _____ + _____

22A $Z = \bar{A}B + A\bar{B}$

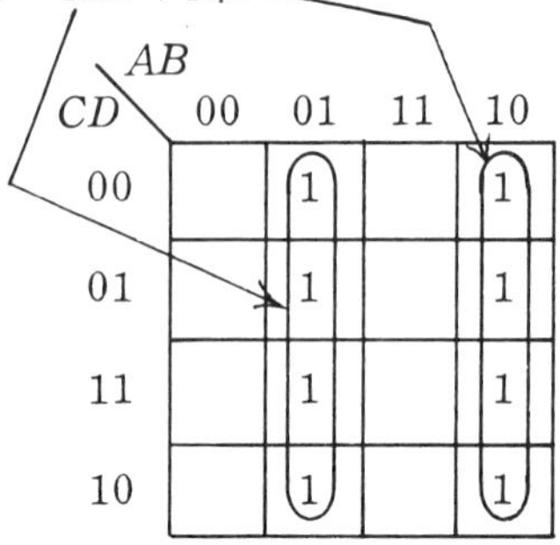

23Q In the case of 2-loops several loops could overlap. The same is true for 4-loops. The two 4-loops on the map below are:

CD \ AB	00	01	11	10
00		1	1	
01		1	1	
11		1		
10		1		

$X =$ ______ + ______

23A $X = \bar{A}B + B\bar{C}$

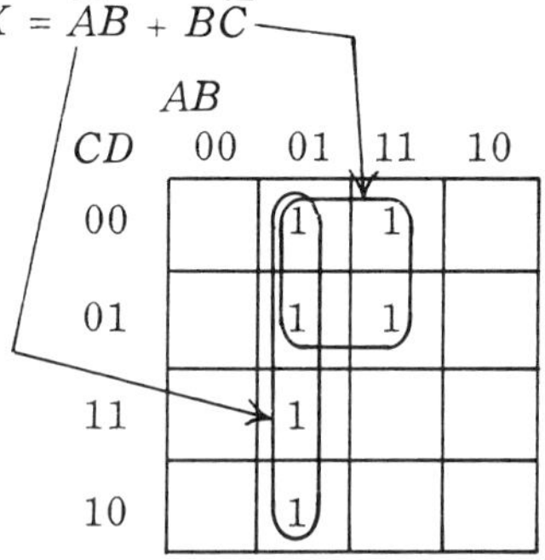

24Q In frame 17 we found that two adjacent 2-loops are equivalent to one 4-loop. Likewise two adjacent 4-loops can be combined to one 8-loop. An 8-loop is described by three less variables than the map. The map below contains two adjacent 4-loops. The 8-loop covering the function is described by the single variable _____.

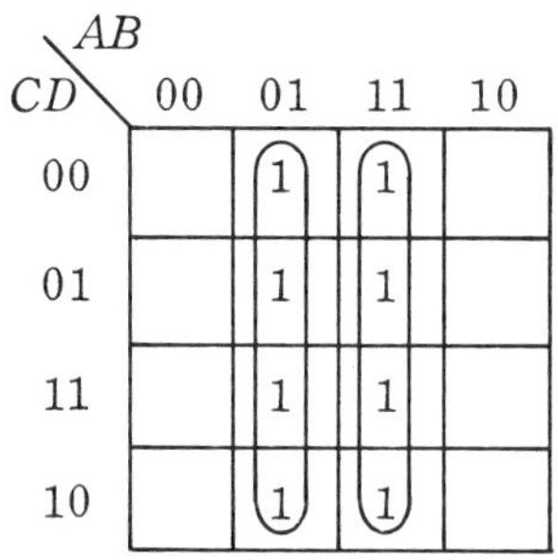

24A *B*

The value of *B* remains constant for all eight 1's.

25

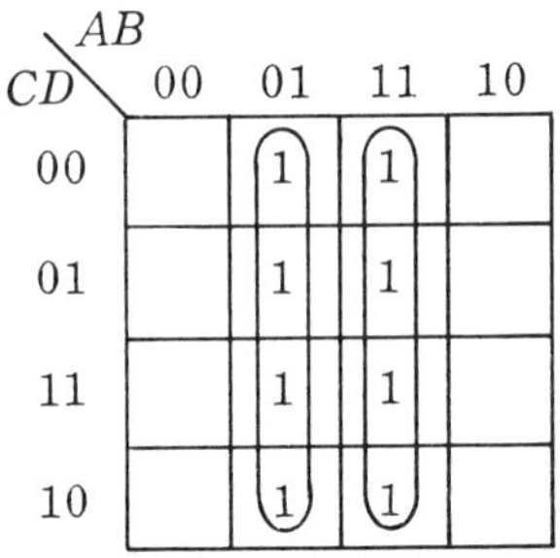

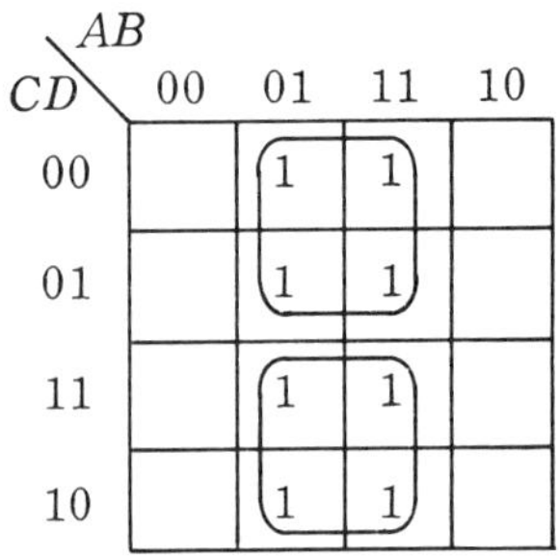

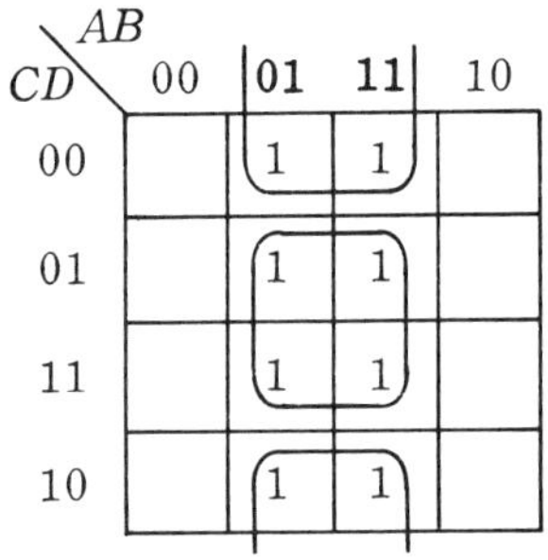

With this function there are three pairs of 4-loop which can combine to the same 8-loop. This corresponds to the three identical entries in the fourth list in the Quine-McCluskey method.

4, 5, 6, 7	01-- ✓	4, 5, 6, 7, 12, 13, 14, 15	-1--
4, 5, 12, 13	-10- ✓	4, 5, 6, 7, 12, 13, 14, 15	-1--
4, 6, 12, 14	-1-0 ✓	4, 5, 6, 7, 12, 13, 14, 15	-1--
5, 7, 13, 15	-1-1 ✓		
6, 7, 14, 15	-11- ✓		
12, 13, 14, 15	11-- ✓		

26 Summary

These are the loop sizes introduced thus far:

N = number of variables for which map is drawn
M = number of variables whose value is the same for all 1's in the loop

1-loop: $M = N$
2-loop: $M = N - 1$
4-loop: $M = N - 2$
8-loop: $M = N - 3$

Some interesting facts are

1. The size of a loop is always a power of two.
   Number of 1's in a 1-loop: $1 = 2^0$
   Number of 1's in a 2-loop: $2 = 2^1$
   Number of 1's in a 4-loop: $4 = 2^2$
   Number of 1's in a 8-loop: $8 = 2^3$

2. Loops can overlap. The number of 1's shared by two loops is always a power of two.

---

27 It is the purpose of the map to minimize an expression. You already know that larger loops are cheaper than small ones because they are described by fewer variables. Thus when choosing loops, the fewest and the largest loops which contain all 1's should be picked. The rule for doing this is as follows:

Rule:

1. Go through the squares one at a time and check if it can be covered by one loop or more loops. If it can be covered in only one way pick that loop and go on. If it can be covered by more than one loop, bypass the square and go on; disregard loops within loops.

2. Cover the remaining 1's by the fewest and largest loops possible.

28Q Example

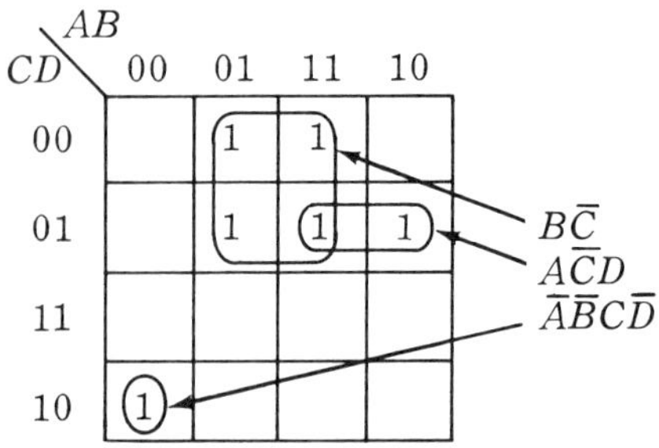

1. There is only one loop covering $\bar{A}\bar{B}C\bar{D}$
2. There is only one loop covering $A\bar{B}\bar{C}D$
3. There is only one loop covering $\bar{A}B\bar{C}\bar{D}$, $\bar{A}B\bar{C}D$, $AB\bar{C}\bar{D}$

$$X = \bar{A}\bar{B}C\bar{D} + A\bar{C}D + B\bar{C}$$

The function Y is plotted on the map below. Read Y using the rule stated in frame 27.

CD \ AB	00	01	11	10
00				
01	1	1		
11	1	1		
10			1	1

$Y =$ ______________

28A

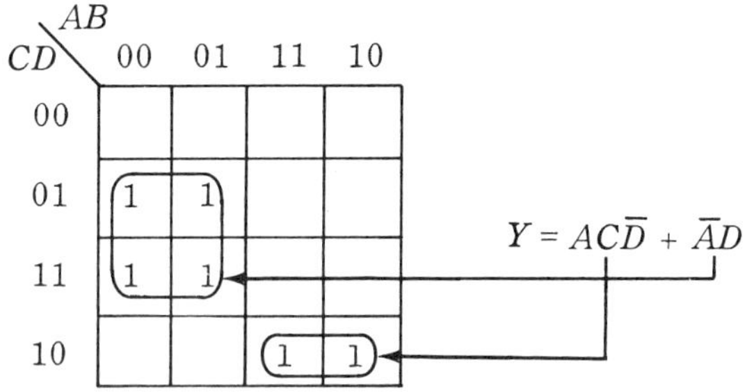

29Q

CD \ AB	00	01	11	10
00				1
01	1	1		
11	1	1	1	1
10				1

$Z =$ ____________________.
Hint: two 4-loops; one 2-loop. Watch for end loops!

29A

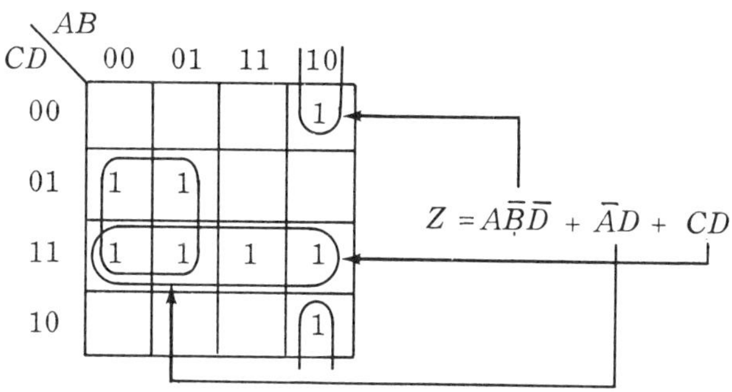

30Q

CD \ AB	00	01	11	10
00	1		1	1
01		1	1	1
11			1	
10	1		1	1

$U =$ ____________________.
Hint: one 2-loop; three 4-loops.

30A

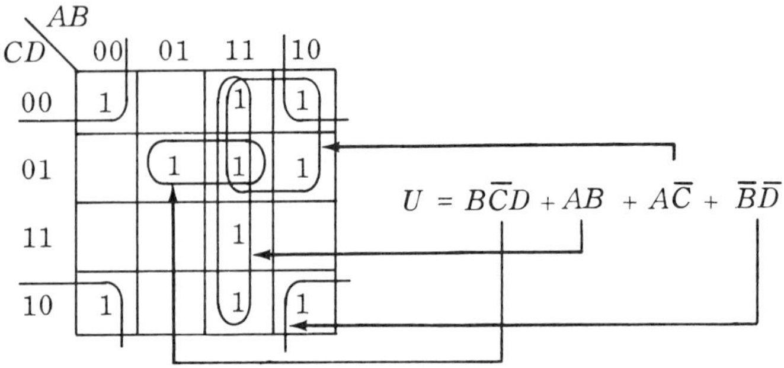

31Q After selecting the 1's which can only be covered by one loop, some 1's may remain unlooped. These remaining 1's are covered by the fewest and largest loops possible. (Rule in frame 27.)

CD \ AB	00	01	11	10
00		1		1
01	1	1	1	1
11	1		1	
10			1	

$V =$ ____________________

31A

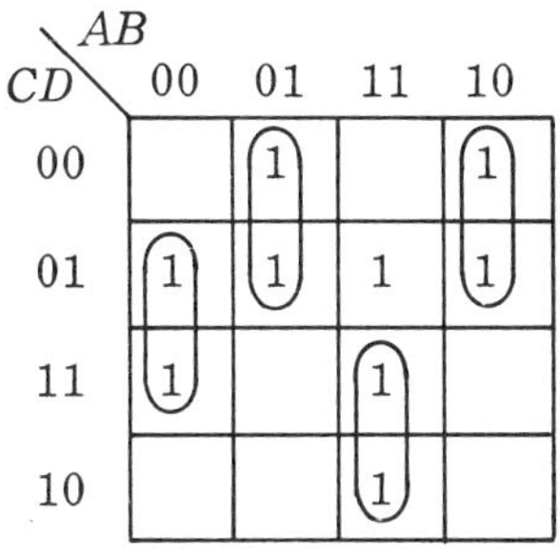

After first pass

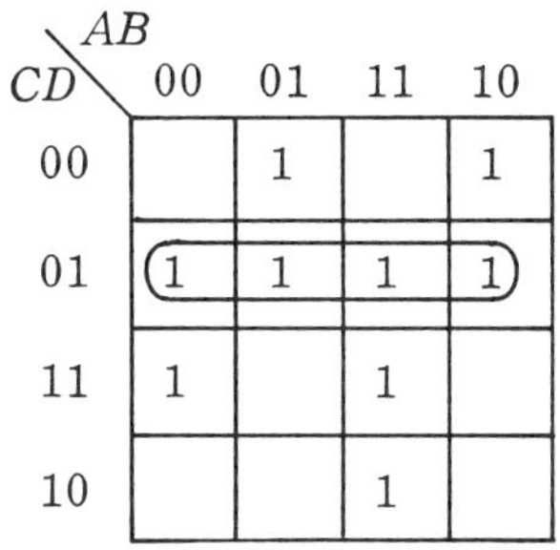

$\bar{C}D$

Correct choice

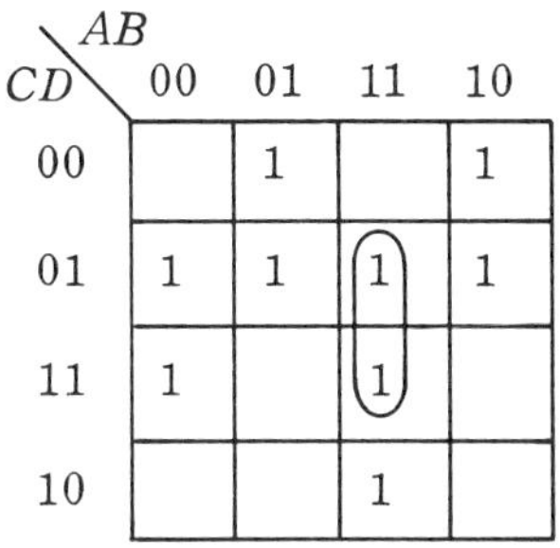

ABD

Incorrect choice

$$V = \bar{A}B\bar{C} + \bar{A}\bar{B}D + ABC + A\bar{B}\bar{C} + \begin{cases} \bar{C}D & \text{(correct choice)} \\ ABD & \text{(incorrect choice)} \end{cases}$$

32Q Read this map (remember end loops):

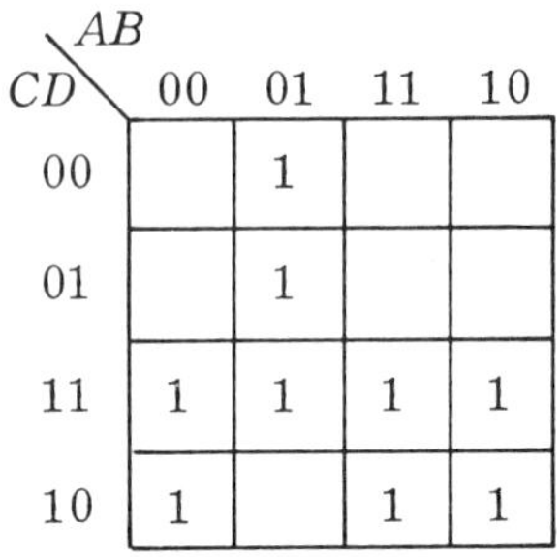

$W =$ ________________

32A $W = \bar{A}B\bar{C} + \bar{B}C + AC + CD$

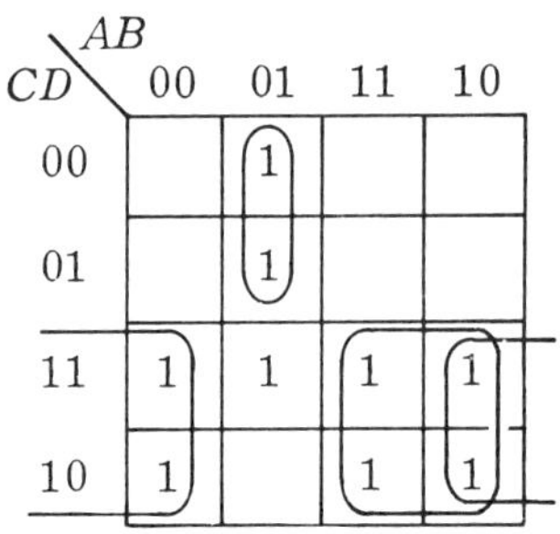

After first pass

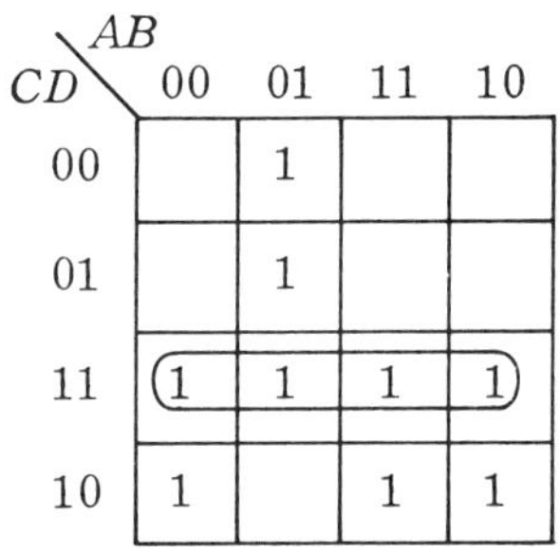

Correct choice
$F = \bar{A}B\bar{C} + \bar{B}C + AC + CD$

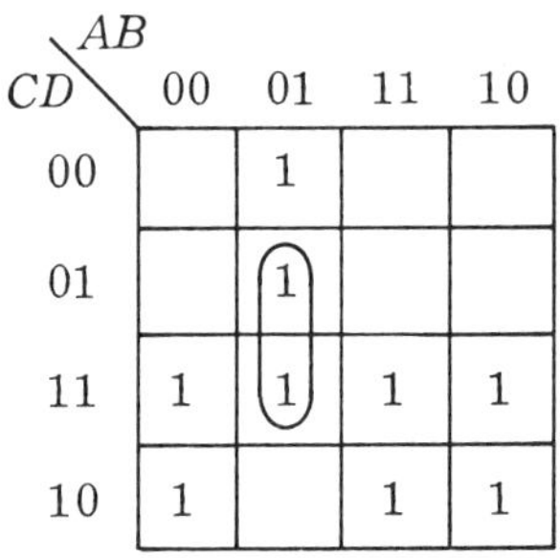

Wrong choice
$F = \bar{A}B\bar{C} + \bar{B}C + AC + \bar{A}BD$

The correct choice has one less variable, thus it is a better solution.

PROBLEMS

Many of the problems given for this chapter are repeats from earlier chapters. This comparison more than anything else demonstrates the speed of solving problems with the Karnaugh map.

1. Simplify the function Z using the Karnaugh map.

$$Z = B\bar{C}\bar{D} + AB\bar{D} + \bar{A}BC\bar{D}$$

(Compare with Prob. 1, Chap. 8, page 150.)

2. Simplify the function Y using the Karnaugh map.

$$Y = A\bar{C}D + AB\bar{C}D + A\bar{B}\bar{C} + \bar{A}BD + BCD$$

3. Repeat (*a*) Prob. 3 and (*b*) Prob. 4 from Chap. 8. Use the Karnaugh map.

4. Plot on a Karnaugh map the function Z from frame 33, Chap. 8.
   (*a*) Show that the entries in the second list correspond to *all* possible 2-loops.
   (*b*) Plot Z on a new map and show that the third list corresponds to *all* possible 4-loops.
   (*c*) Plot Z on a new map and show that the fourth list corresponds to *all* possible 8-loops.
   (*d*) Show that checked terms in the third list correspond to 4-loops which were completely contained in larger loops. (This process could be done for all checked terms.)

5. (*a*) Find the function Z plotted on the map.
   (*b*) Find a 2-loop not used in the answer.
   (*c*) Compare this redundant loop with the redundant prime implicant, frame 14A, Chap. 9. (The prime-implicant chart in frame 14A corresponds to the map in this problem.)

CD \ AB	00	01	11	10
00		1		
01	1	1	1	1
11			1	
10			1	

6. Why is prime implicant *BCD* in Chap. 9, frame 21A, redundant? (Show reasoning on the Karnaugh map.)

7. Repeat Prob. 5, Chap. 9, using the Karnaugh map.

8. The Karnaugh map is only one graphic format for displaying a function. Another popular form is the Veitch diagram shown here. Instead of binary numbers heading the columns and rows, they are headed by the actual variables. In each of the squares write the standard sum term in decimal notation which corresponds to that square (A is high order as always).

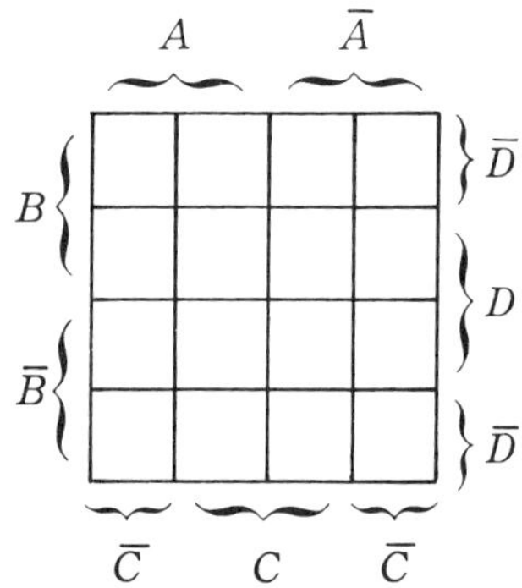

9. Read the function *X* from this Veitch diagram. Loop just like a Karnaugh map but watch the column headings.

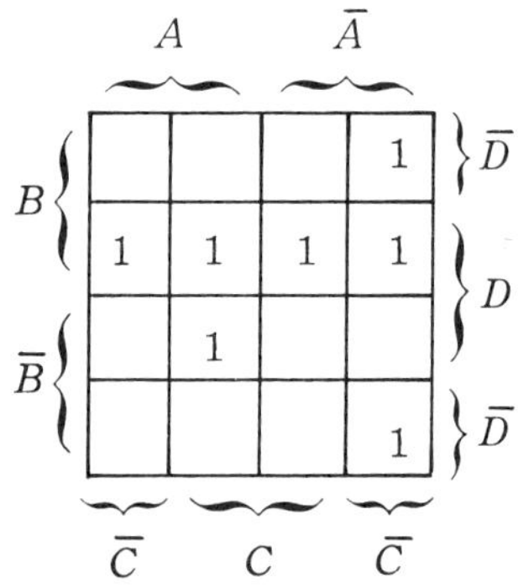

Chapter 12

KARNAUGH MAP—III

The first part of this chapter shows how the inverse of a function is read from a Karnaugh map. The second part introduces a *don't care* condition (an input condition for which it does not matter what the output is) and shows how to plot and how to take advantage of these conditions on a Karnaugh map.

1 The first topic to be covered in this chapter is the inversion of a function using the Karnaugh map. Let us go back and compare the map with a truth table.

A	B	X
0	0	1
0	1	0
1	0	0
1	1	1

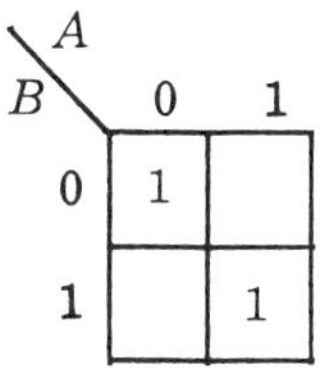

X

To obtain $\bar{X}$ from X, the entry for X in each row is inverted separately (1→0, 0→1). Turning to the map, each square in the map corresponds to one row in the table. Thus to obtain a map for $\bar{X}$ from a map for X, each square is changed from 0 to 1 or from 1 to 0, remembering that the blank squares should contain 0's. The map for $\bar{X}$ therefore has two 1's:

A	B	X	$\bar{X}$
0	0	1	0
0	1	0	1
1	0	0	1
1	1	1	0

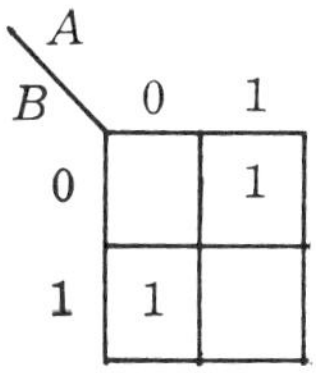

Map for $\bar{X}$

Reading $\overline{X}$ from the map, it is $\overline{X} = A\overline{B} + \overline{A}B$, which agrees with the truth table.

Rule: Given a map for a function Y, the map for $\overline{Y}$ can be found by replacing all 1's by 0's and all 0's by 1's.

2Q Using the rule for obtaining the inverse of a function on the map, fill in the map for $\overline{Z}$.

BC \ A	0	1
00	1	
01	1	1
11		1
10	1	

Z

BC \ A	0	1
00		
01		
11		
10		

$\overline{Z}$

2A

BC \ A	0	1
00		1
01		
11	1	
10		1

The 0's (blank's) are changed to 1's.

3Q

CD \ AB	00	01	11	10
00	1	1	1	
01		1	1	1
11		1	1	1
10	1	1	1	

U

CD \ AB	00	01	11	10
00				
01				
11				
10				

$\overline{U}$

Fill in the map for $\overline{U}$ and write an equation for $\overline{U}$. U = ____________________ . (Be careful when reading the map. Always read the largest possible loops.)

3A

CD \ AB	00	01	11	10
00				1
01	1			
11	1			
10				1

$\overline{U} = A\overline{B}\overline{D} + \overline{A}\overline{B}D$

4Q Try another example:

CD \ AB	00	01	11	10
00	1	1	1	1
01		1	1	
11		1	1	
10				

V

CD \ AB	00	01	11	10
00				
01				
11				
10				

$\overline{V}$

$\overline{V}$ = ____________________

4A

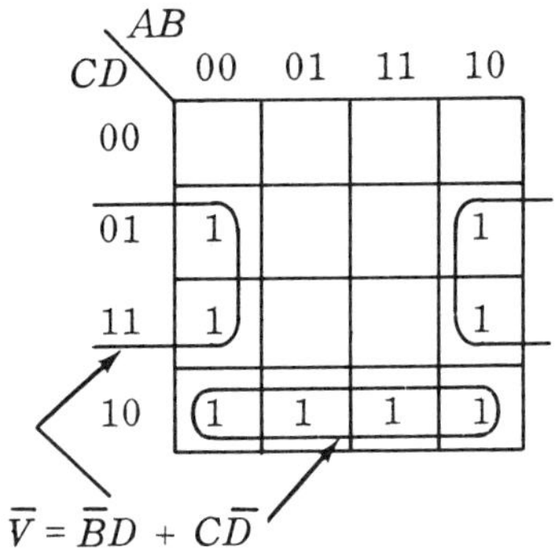

5Q One is not required to redraw the map for the inverse and then read the inverse. The inverse can be found directly by reading the 0's of the original map. For example,

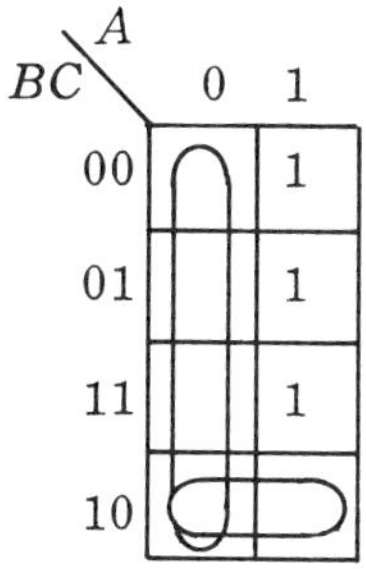

If W is plotted on this map $\overline{W}$ is:

$\overline{W} = \overline{A} + B\overline{C}$

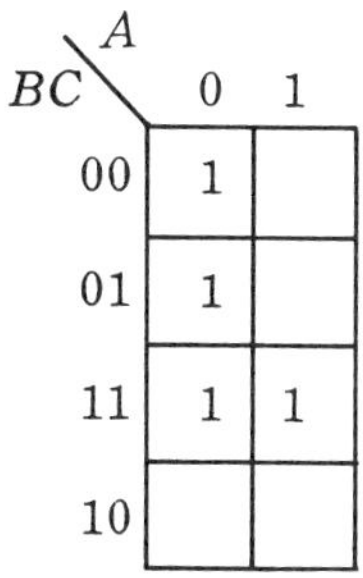

If X is plotted on the map on the left:

$\overline{X} =$ __________

5A

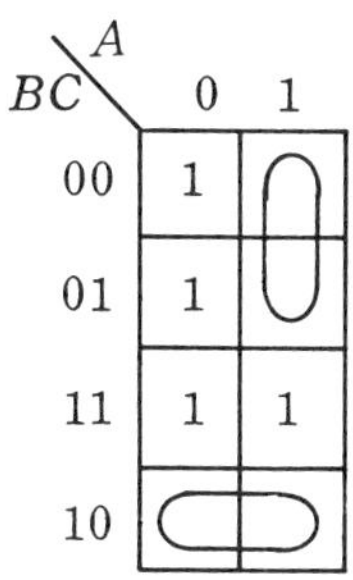

$\overline{X} = A\overline{B} + B\overline{C}$

---

6 **Review**

The inverse of a function is read from the map by using one of the following methods: (1) drawing a new map, plotting the inverse of the function on a new map (interchanging 0's and 1's), and then reading the 1's of the new map; or (2) looping and reading the 0's of the original map.

7Q In the last example, $\overline{X}$ was found to be:

$$\overline{X} = A\overline{B} + B\overline{C}$$

What would we get if we were to invert this equation? Do *not* multiply out.

$$X = \underline{\hspace{5cm}}$$

7A

$$X = (\overline{A} + B)(\overline{B} + C)$$

$$X = (\overline{A\overline{B}})(\overline{B\overline{C}})$$

$$= (\overline{A} + B)(\overline{B} + C) \qquad \text{[answer]}$$

---

8 The result in frame 7A is very interesting because it shows an expression in product form and at the same time it is minimal. Without going any further we can see that there are really *two* minimal forms for *every* expression. One is a sum form, the other a product form. The minimal sum form can be obtained directly from the map. But to find the minimal product form of an expression, the inverse of that expression must be found. Then this inverted expression is minimized (minimal sum form of inverse). The final step is to reinvert the minimized expression. Sometimes the product form is cheaper to implement, sometimes the sum form is cheaper. In a practical design both should be tried.

---

9 **Don't Cares**

In some problems the inputs are interdependent in such a way that certain input combinations cannot occur.

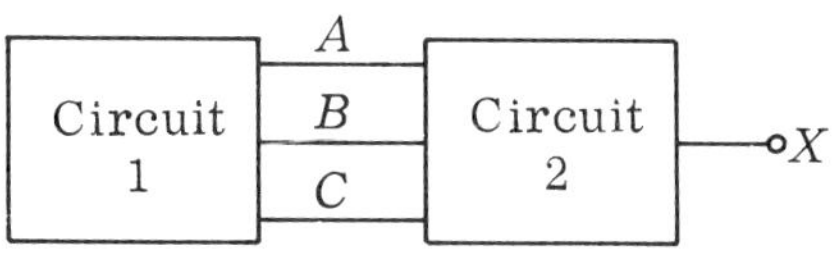

Suppose the variables A, B, and C are interrelated in circuit 1. When designing circuit 2 this interdependency can be made use of. Although some input combinations are not possible (due to circuit 1) the output of circuit 2 has to be considered for *all* possible input combinations. From this example we can conclude that for some input combinations the output X has to be considered, although we know that the particular combination of A, B, and C will never occur.

For these conditions we "don't care" about the output; thus we define the output as a *don't care*, meaning it can be either 0 or 1. Another source of don't cares is when the output of a circuit is not used when certain input conditions occur.

When plotting don't cares on the map, the squares cannot be left blank because this would denote a 0. To show a don't care, we use a special symbol consisting of a 0 superimposed on a 1 (Ø).

10Q Assume a three-variable problem with the variables A, B, and C. These are interrelated in such a way that no more than two of them can be equal to 1 at any given time. All three can never be equal to 1 simultaneously. Which square(s) in the map can be considered don't cares?

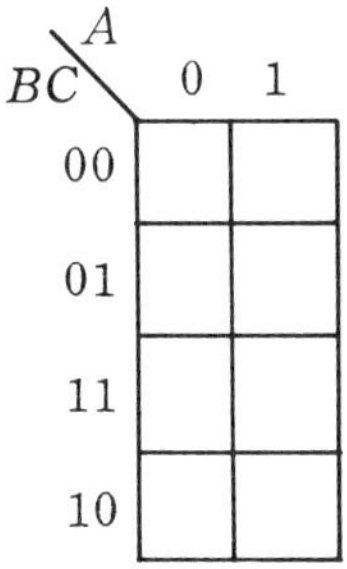

10A

BC \ A	0	1
00		
01		
11		Ø
10		

11Q In this problem the output function is 1 whenever exactly two of the four inputs are 1, and we *don't care* what the output is when three or all four inputs are 1. The output is 0 whenever none of the inputs is 1 or if only one of the inputs is 1. Plot this function on the map.

CD \ AB	00	01	11	10
00				
01				
11				
10				

11A

CD \ AB	00	01	11	10
00			1	
01		1	Ø	1
11	1	Ø	Ø	Ø
10		1	Ø	1

12 How do we obtain the most minimal circuit if the map contains don't cares? Don't care conditions are chosen to be either 0 or 1, depending on which choice results in a more minimal circuit, which means the largest loops. However, we will not try to loop as many or as few *don't cares* as possible.

Examples

Assume a function X equals $\bar{A}BD$ with don't care conditions as shown on the right.

CD \ AB	00	01	11	10
00				∅
01		1	∅	
11		1	∅	
10				

Two don't cares are utilized to reduce X.

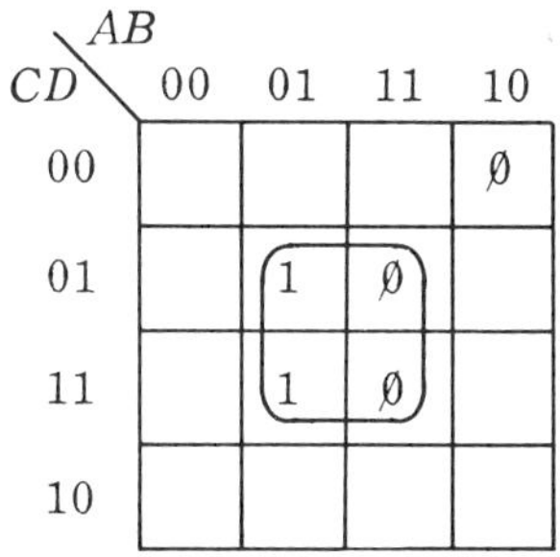

$X = BD$

Two don't cares are made equal to 1 to enlarge the original loop. One don't care is made 0 because it does not enlarge any loop.

13Q Loops which contain only don't cares should not be included in the answer.

Example

CD \ AB	00	01	11	10
00		1		
01		1		
11			∅	∅
10			∅	∅

Assuming all don't cares are 0:

$$X = \overline{A}B\overline{C}$$

Assuming all don't cares are 1:

$$X = \overline{A}B\overline{C} + AC$$

The second form of X contains a loop of only don't cares, which is not necessary because all the don't cares could have been assumed to be 0's.

CD \ AB	00	01	11	10
00		∅	1	
01			1	
11			∅	∅
10			∅	∅

$Y =$ __________

13A $Y = AB$

CD \ AB	00	01	11	10
00		∅	1	
01			1	
11			∅	∅
10			∅	∅

14Q The general rule for use of don't cares is as follows.

Rule: For each 1 in the map try to make the largest loop. The loop may contain either other 1's or don't cares. Then choose the fewest loops to cover all the 1's. Don't care conditions inside the chosen loops are made 1, the others are made 0.

Read the expression Z plotted on the map below.

$CD \backslash AB$	00	01	11	10
00			1	
01		∅	∅	∅
11	1	∅	1	∅
10			∅	

$Z =$ ________________

14A $Z = AB + CD$

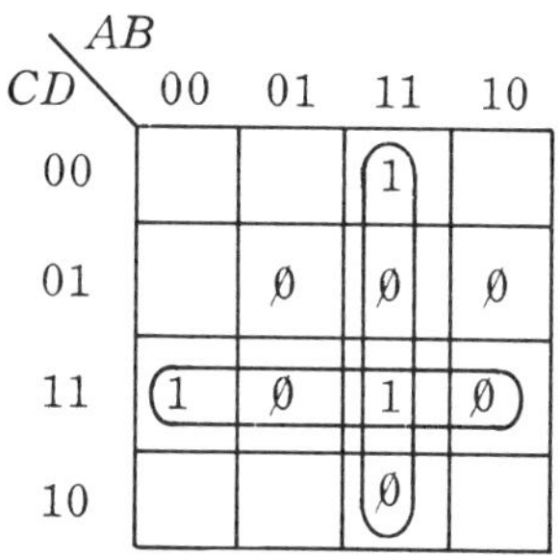

Four don't cares were chosen to be 1, two were chosen to be 0.

15Q

CD \ AB	00	01	11	10
00		∅	∅	
01			1	1
11		∅	∅	
10			1	1

$W =$ ____________

15A $W = A\overline{C}D + AC\overline{D}$

CD \ AB	00	01	11	10
00		∅	∅	
01			1	1
11		∅	∅	
10			1	1

The don't cares did not help at all in this problem.

---

16 How do we find the inverse of a function if it contains don't cares? By definition, we do not care what the value of X is for a don't care condition. Thus we do not care what the value of $\overline{X}$ is for that same condition. If we are working with $\overline{X}$ we choose the don't care conditions to give the simplest expression for $\overline{X}$.

Rule: When deriving the inverse of an expression: 1's go to 0's; 0's go to 1's; ∅'s remain ∅'s.

Example

CD \ AB	00	01	11	10
00	1	∅		1
01	1	∅	∅	
11			1	1
10			1	

Y

CD \ AB	00	01	11	10
00		∅	1	
01		∅	∅	1
11	1	1		
10	1	1		1

$\overline{Y}$

17Q When looping the inverse of a function the same rule applies that applied for looping the function itself.

CD \ AB	00	01	11	10
00			∅	
01		1	1	
11	∅	∅	1	
10				

V

Plot $\overline{V}$ and read it from the map (loop carefully).

CD \ AB	00	01	11	10
00				
01				
11				
10				

$\overline{V}$

$\overline{V}$ = ________________

17A $\overline{V} = \overline{B} + \overline{D}$

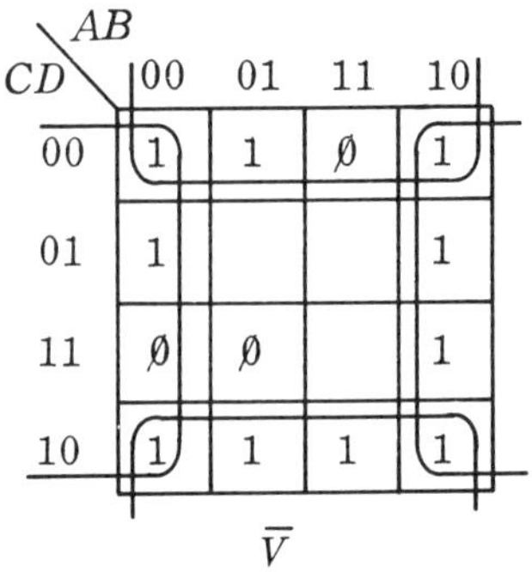

18Q Try this problem without replotting:

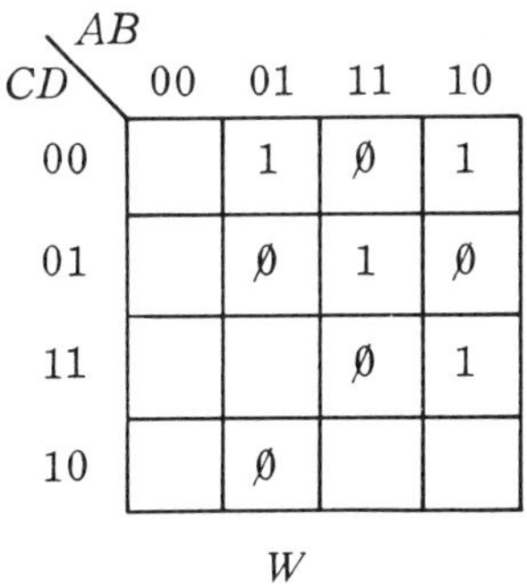

$\overline{W}$ = ____________________

18A These are the three equally good answers:

$\overline{W} = \overline{A}\,\overline{B} + \overline{A}C + C\overline{D}$

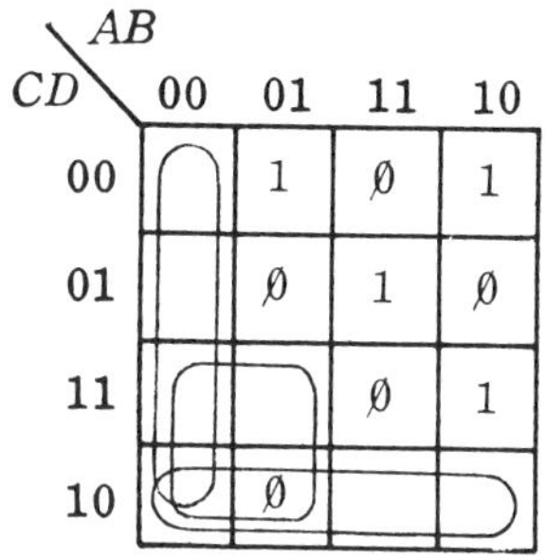

$\overline{W} = \overline{A}\,\overline{B} + \overline{A}D + C\overline{D}$

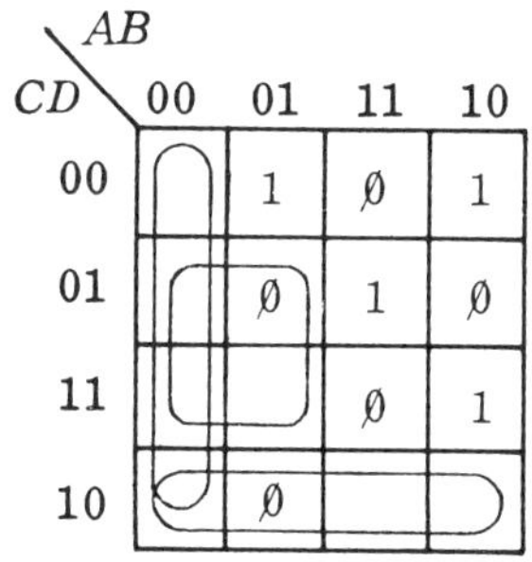

$\overline{W} = \overline{A}\,\overline{B} + BC + C\overline{D}$

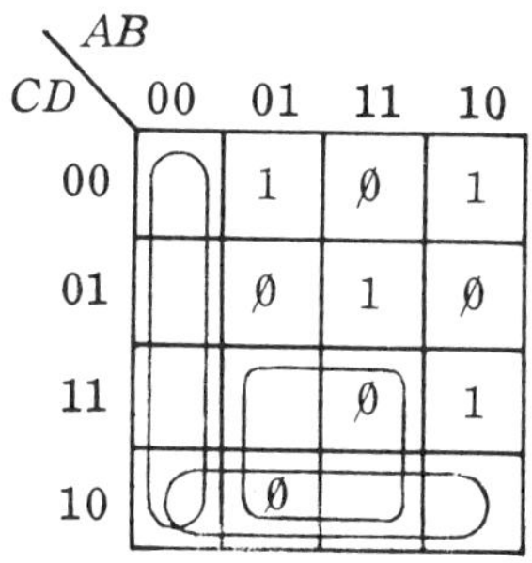

19 Summary

Don't cares are introduced in problems either because some input conditions cannot occur or because for some input conditions the output will not be used.

These don't cares can be chosen to be either 0 or 1, depending upon which makes the more minimal circuit (larger loops). The criteria for looping is that all 1's must be covered by loops, 0's must *not* be covered by loops, and the don't cares may or may not be covered.

When inverting an expression on a map, the don't cares remain in the same places.

When looping the 0's of a map with don't cares, the same rule applies which applied for the 1's. The don't cares may, but need not, be enclosed in the loops.

The concept of don't care conditions was introduced in conjunction with the maps. The same concept can be incorporated in other minimization techniques.

PROBLEMS

1. Obtain the minimum product form expression for:

$$X(A, B, C, D) = \Sigma(1, 3, 4, 6, 11, 12, 13)$$

2. If both the true and inverted values of the outputs are required, determine whether it would be better to form the true form first and then invert it, or to first generate the inverted form and invert it. (As criteria use the number of diodes required in the implementations. Do not consider transistors.)

   (*a*) $X(A, B, C, D) = \Sigma(3, 5, 11, 12, 13, 15)$
   (*b*) $Y(A, B, C, D) = \Sigma(1, 3, 4, 6, 8, 9, 11)$
   (*c*) $Z(A, B, C, D) = \Sigma(0, 1, 3, 7, 8, 9, 10, 11, 12, 14)$

3. Assuming only true outputs are required, for the three functions given in Prob. 2, would it be better to implement the output in product or in sum form? (*a*) X; (*b*) Y; (*c*) Z.

4. Find an equivalent but simpler circuit for the one below:

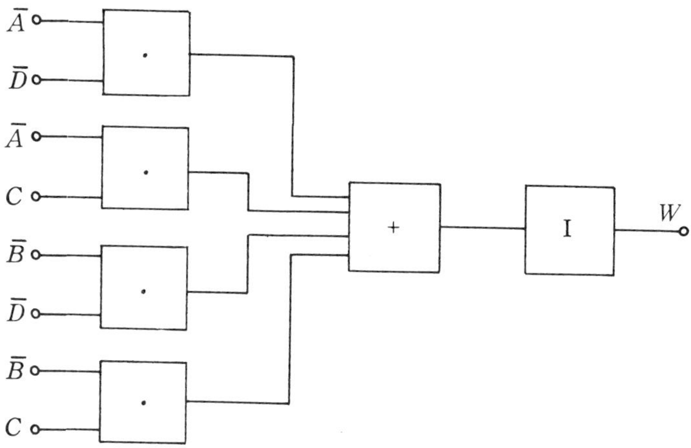

5. Four binary lines are required to represent a decimal digit. Since the four binary lines have 16 possible values, six of the values cannot be used. Assuming that the decimal digits 0 to 9 are given by the four-digit binary count of 0000 to 1001, and the counts 1010 to 1111 never occur, design the best circuit whose single output will be 1 if the inputs on the four inputs lines (whose inverse is available) represent (*a*) 9 and (*b*) 6.

6. Subtraction is often performed in a decimal computer by a process known as 9's complement addition. The 9's complement of a decimal digit (m) is equal to ($9 - m$). Assume that a decimal digit is represented by a four-digit binary number and that the values of 10 to 15 can never occur. Design four separate circuits having four binary inputs such that the combined value of the outputs is the decimal digit which is the 9's complement of the input digit. (*Hint:* Draw a truth table for the inputs A, B, C, D and outputs W, X, Y, Z. Simplify and implement first W, then X, etc. Treat the four circuits as four separate problems.)

7. The don't cares can also be made use of in the Quine-McCluskey method. Here the don't cares are added to the list of 1's and the combining of terms into higher-order lists proceeds as if there were only 1's. This means that all possible terms of higher-order lists are generated. When the prime implicants are selected in the prime-implicant chart, only those terms are listed at the top of the chart which must be included in the answer.

This means that only the terms corresponding to the 1's are listed along the top. Along the side all prime implicants are listed. The selection of prime implicants is the same as always. Simplify the following functions (assume the variables A, B, C, and D):

(*a*) $X = 1$ for these terms: 1, 5, 11, 15
$X = \emptyset$ for these terms: 7, 9, 12, 13

(*b*) $Y = 1$ for these terms: 4, 5, 14, 15
$Y = \emptyset$ for these terms: 7, 13

Chapter 13

ADVANCED TOPICS

This chapter discusses three separate topics:

1. Multiple Outputs
   A method is given for sharing logic blocks between two or more circuits which have common inputs.

2. Factoring
   Introduces the concept of designing circuits having more than two logic levels.

3. Implementing with other logic blocks.
   Describes other logic blocks, and gives examples for designing logic circuits with NAND blocks.

Multiple Outputs

In digital computers, it is very common to find that two or more output functions use the same input variables. It is also common to find two or more output functions which share some, but not all, of their inputs. Although these circuits can be designed separately, it is often possible to reduce the amount of circuitry by treating all the functions as part of the same problem. The resulting circuit is called a *multiple-output circuit*. A very simple but effective multiple-output method can be used when two or more expressions contain the same term. The term can be implemented once and used in both expressions.

Example

The functions X and Y are both functions of the same variables. Implement them together, trying to share blocks. Assume all variables are available in both true and inverted form.

$X = \bar{A}\bar{B} + \bar{C}D + \bar{A}\bar{C}$
$Y = \bar{A}\bar{B}\bar{C} + \bar{A}BD + \bar{C}D$
Implemented separately,

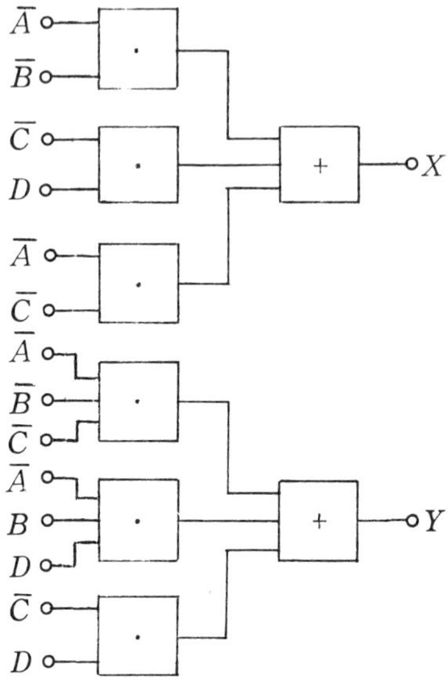

In the foregoing, it can be seen easily that $\bar{C}D$ is generated twice. Two diodes can be saved if one AND generating $\bar{C}D$ is eliminated and the other fed into both ORs. The new circuit is

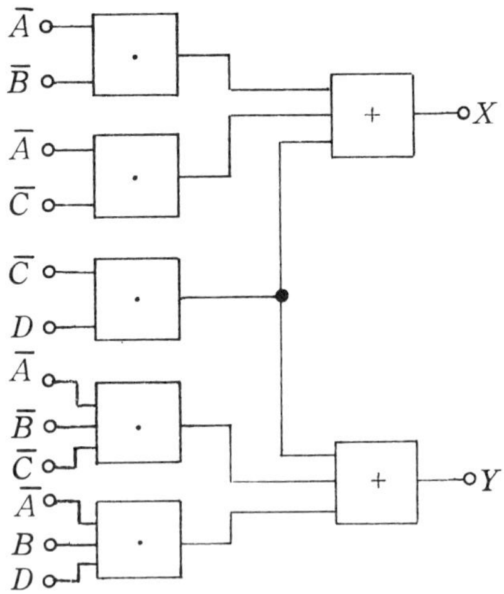

You have become familiar with maps in the last few chapters and probably realize that each loop on the map corresponds to one AND circuit. Thus if we want to use these ANDs for two outputs, the same loops have to appear on both maps. Returning to the example used above:

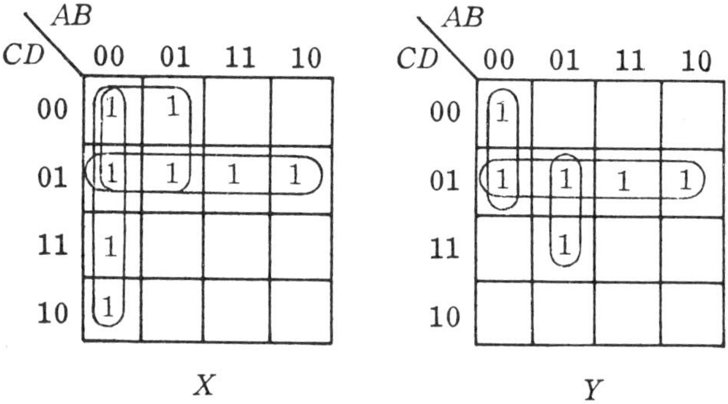

The looping for X and Y demonstrates that there is one loop ($\bar{C}D$) common to both. Thus two diodes (one AND having two inputs) can be saved.

It has been pointed out that if terms are to be shared the same loops have to appear on the corresponding maps. Consider the following three maps with the outputs X, Y, and Z.

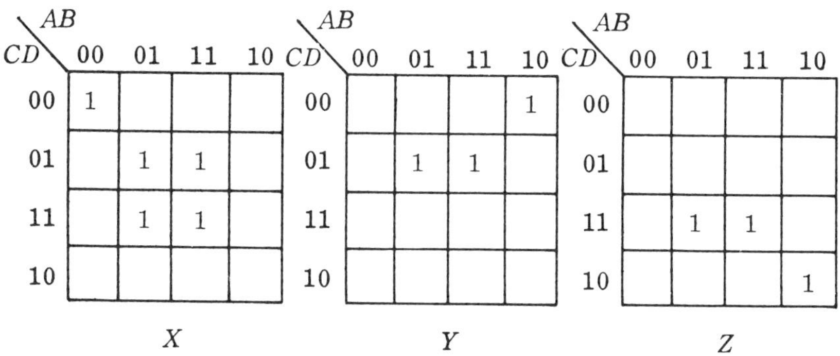

If each map were looped independently there would be no common term. The conventional looping is indicated on the set of maps on page 240.

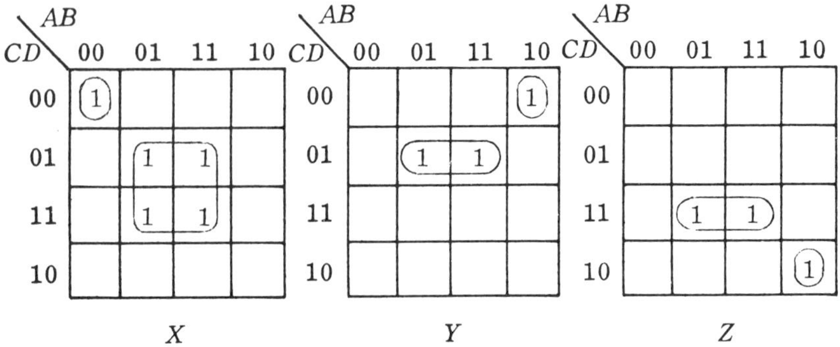

But if, instead of looping the map for X in a 4-loop and one 1-loop (as shown above), it were looped with one 1-loop and two 2-loops (as shown below), these two 2-loops could be shared with Y and Z.

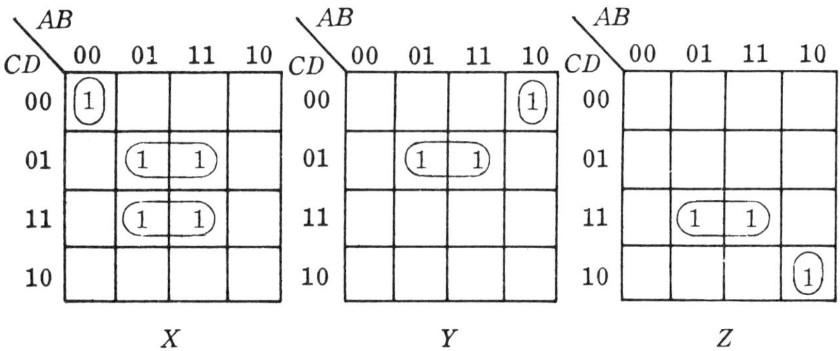

The interesting point to be learned from this example is that under special conditions the largest loop is not used.

Rule: A large loop is broken down into smaller ones only if each square in the large loop can be covered by loops already existing in other functions.

Let us try some examples of increasing complexity.

1Q Implement X, Y, and Z in 14 diodes and 2 transistors. Assume that the variables are only available in their true form.

CD \ AB	00	01	11	10
00			1	1
01			1	
11			1	
10			1	

X

CD \ AB	00	01	11	10
00			1	1
01		1	1	
11				
10				

Y

CD \ AB	00	01	11	10
00			1	
01		1	1	
11			1	
10			1	

Z

1A

CD \ AB	00	01	11	10
00			1	1
01			1	
11			1	
10			1	

$X = A\overline{C}\overline{D} + AB$

CD \ AB	00	01	11	10
00			1	1
01		1	1	
11				
10				

$Y = A\overline{C}\overline{D} + B\overline{C}D$

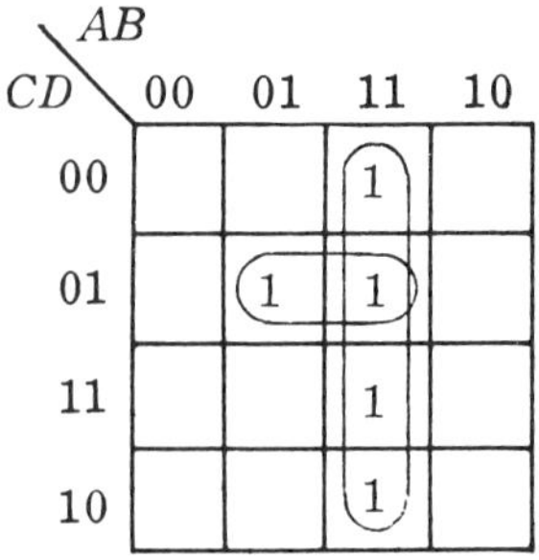

$Z = AB + B\overline{C}D$

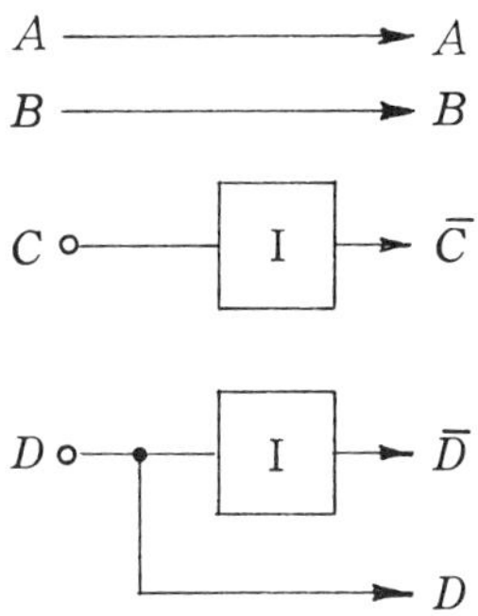

Circuit for X, Y, and Z

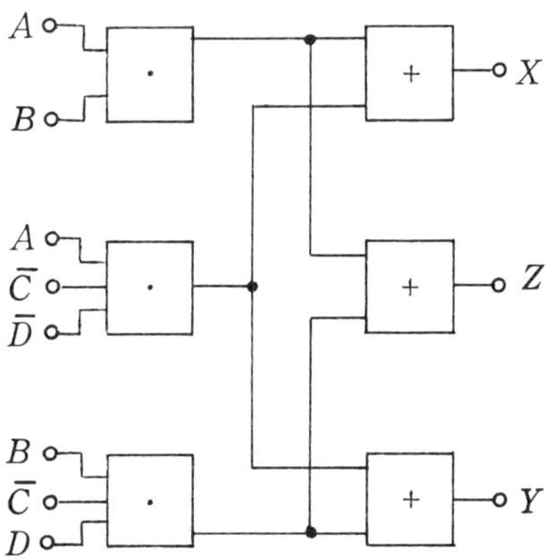

2Q Implement these three outputs, assuming that inverted variables are not available.

CD \ AB	00	01	11	10
00				
01	1	1	1	1
11				1
10				1

X

CD \ AB	00	01	11	10
00			1	1
01	1	1	1	
11		1		
10		1		

Y

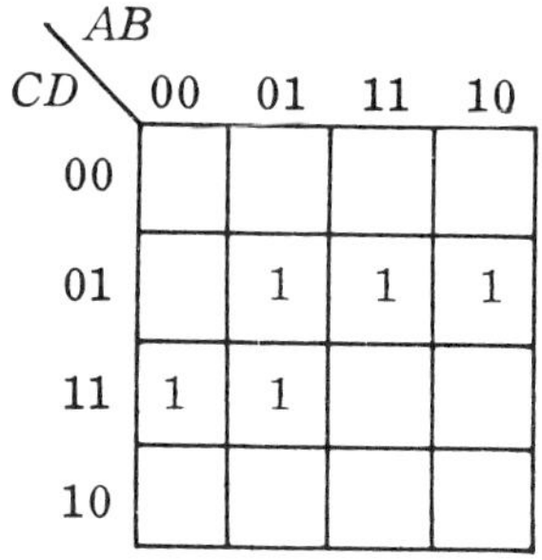

Z

2A

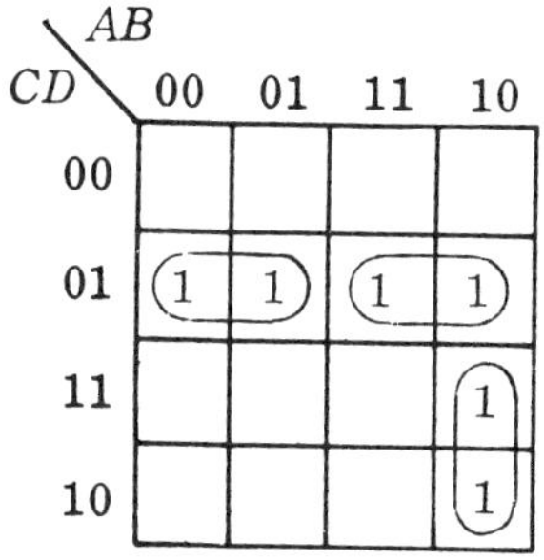

X

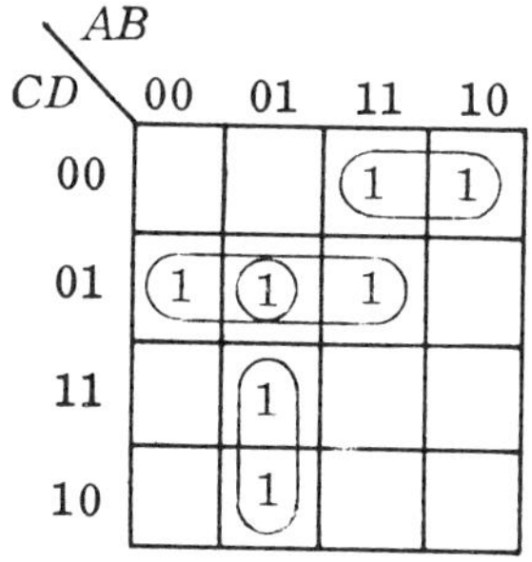

Y

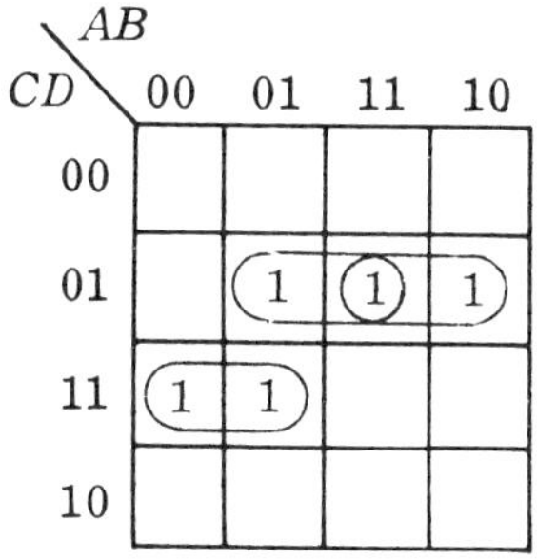

Z

$$X = \bar{A}\bar{C}D + A\bar{C}D + A\bar{B}C$$
$$Y = A\bar{C}\bar{D} + \bar{A}\bar{C}D + B\bar{C}D + \bar{A}BC$$
$$Z = B\bar{C}D + A\bar{C}D + \bar{A}CD$$

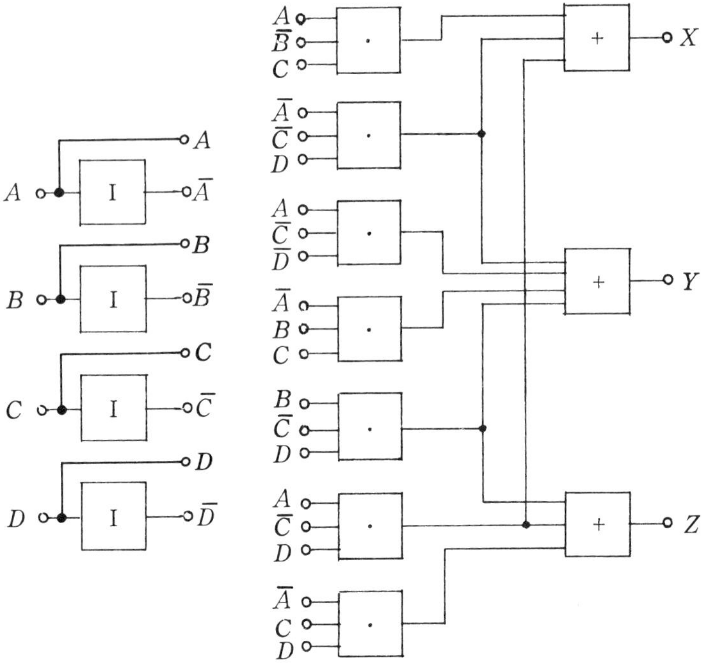

3Q Implement X and Y, assuming that all variables are available in both true and inverted form.

CD \ AB	00	01	11	10
00				
01		1	1	1
11		Ø	Ø	
10				

X

CD \ AB	00	01	11	10
00		Ø	1	1
01		Ø		
11		1	Ø	
10			Ø	

Y

3A

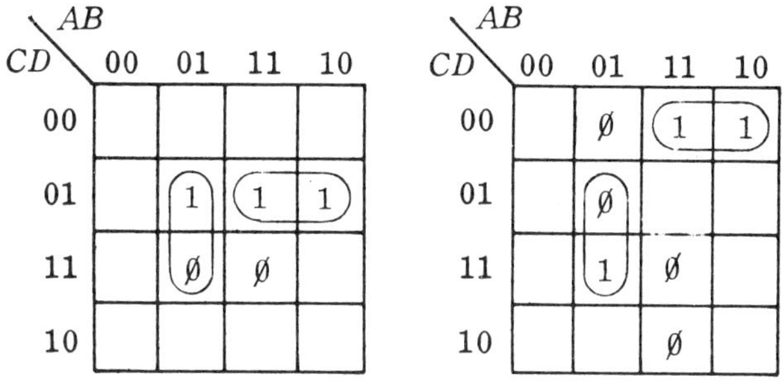

$X = A\bar{C}D + \bar{A}BD$ $\qquad$ $Y = \bar{A}BD + A\bar{C}\bar{D}$

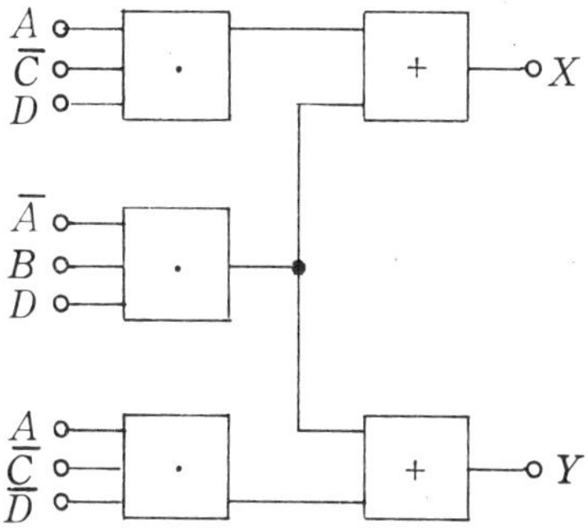

13 diodes

Factoring

In our studies thus far, we have obtained answers in the form of ANDs followed by OR and ORs followed by AND. These answers are referred to as two-level answers because the signal has to travel through two levels (one level of AND circuits and one level of OR circuits) to go from the input to the output. If variables are not available in both true and inverted form, INVERTERS may be needed on the inputs, but these INVERTERS are not counted as a level.

Generally, factoring will extend the number of levels from two to three (as for example O-A-O or A-O-A). The factoring theorem is:

$$AB + AC = A(B + C)$$

You may note that the factoring theorem for Boolean algebra is the same as for conventional algebra, namely, that common factors are taken out. Some factoring examples are:

1. $AB\overline{C} + A\overline{B}D = A(B\overline{C} + \overline{B}D)$
2. $AB\overline{C} + \overline{A}B\overline{C} = B\overline{C}(A + \overline{A})$
3. $A\overline{B}\overline{C} + AD + \overline{B}CD = A(\overline{B}\overline{C} + D) + \overline{B}CD = \overline{B}(A\overline{C} + CD) + AD =$
   $D(A + \overline{B}C) + A\overline{B}\overline{C}$

There are no simple rules for using factoring techniques to minimize a circuit. Factoring sometimes reduces the cost of implementing a circuit, but not always. This fact will be demonstrated by some examples.

1Q Implement the function X without factoring it. Assume that all variables are available in both true and inverted form.

$$X = ACD + A\overline{B}C + AC\overline{E}$$

The circuit requires ______ diodes.

1A Twelve diodes

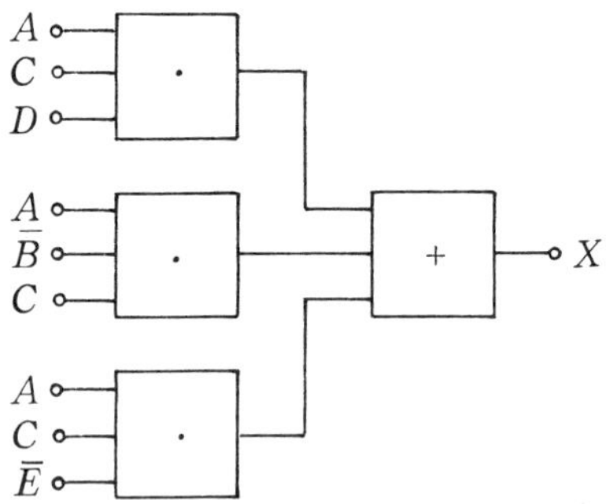

2Q The same expression,

$$X = ACD + A\bar{B}C + AC\bar{E}$$

implemented in factored form, requires ______ diodes.

2A Six diodes

$$X = AC(D + \bar{B} + \bar{E})$$

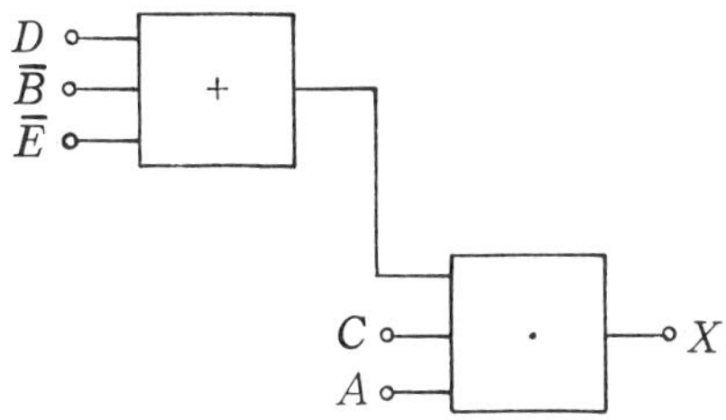

3 Factoring reduced the cost of implementation in frame 2A.

4Q Implementing Y without factoring requires ______ diodes.

$$Y = A\bar{B}C + AB\bar{C} + D$$

(Assume that all variables are available in both true and inverted form.)

4A Nine diodes

5Q Implementing Y in factored form requires ______ diodes.

$$Y = A\bar{B}C + AB\bar{C} + D$$

5A Ten diodes

$$Y = A(\overline{B}C + B\overline{C}) + D$$

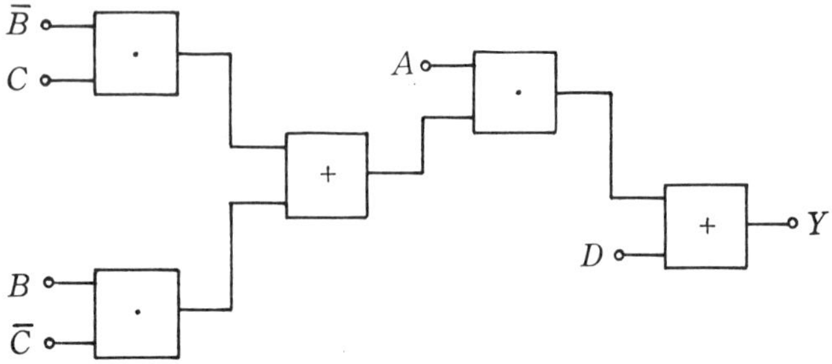

6 The last example should be implemented in the unfactored form, because the factored form is more costly. In conclusion, it can be said that some functions are better implemented in factored form, some in unfactored form. Both have to be tried before deciding.

Implementing with Other Logic Blocks

In Chap. 2 the three most commonly used logic blocks were introduced. They were the AND, the OR, and the INVERTER. These three are by no means the only ones used in logic design.

A list of some common logic blocks, along with their definitions is given in the list below. For the sake of completeness, the AND, OR, and INVERTER are repeated. In the definitions, those conditions are given for which the output is 1. It is understood that the remaining conditions produce a 0 output.

AND	All inputs 1
OR	At least one input 1
INVERTER	Has only one input; output 1 if input 0

EXCLUSIVE OR	Has two inputs; output 1 if either input, but not both, is 1
MAJORITY	Has odd number of inputs; output is 1 if the majority of inputs (more than half) are 1
MINORITY	Has odd number of inputs; output is 1 if a minority of inputs (less than half) are 1
NOR	All inputs are 0
NAND	At least one input is 0
IMPLICATION	Two inputs to be called A and B; output is 1 except when A is 1 and B is 0

Throughout the book, we have made implicit use of the fact that any function can be implemented with AND, OR, and INVERTER blocks. It is also true that other blocks, or sets of blocks, can be used to implement any function. A set of blocks able to implement any function is called a *functionally complete* set. One method of testing a set of blocks for functional completeness is to try to implement the AND, OR, and INVERT functions using only the blocks of the set. Since we know that the AND, OR, and INVERT are functionally complete, then any set that can implement these functions must also be functionally complete. For example, let us test the set made up only of AND and INVERT blocks for functional completeness. If the OR function can be generated with these two blocks, then the group AND-INVERTER is functionally complete.

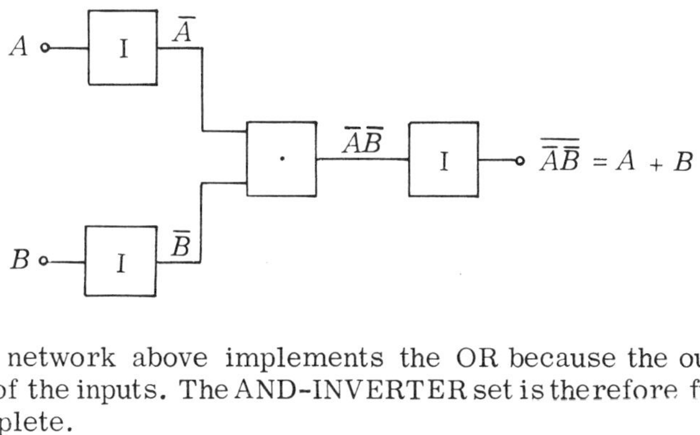

The network above implements the OR because the output is the OR of the inputs. The AND-INVERTER set is therefore functionally complete.

Some blocks are functionally complete by themselves; one of these is the NAND block, which is an AND followed by an INVERTER. A circuit analysis of a single input NAND block shows that it acts as a single input AND followed by an INVERTER; the AND can be neglected; thus it is an INVERTER.

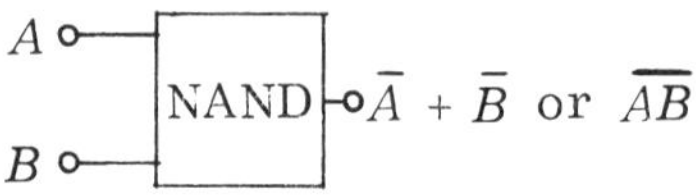

NAND block

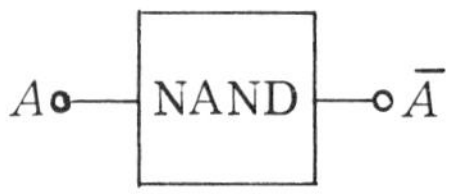

INVERT

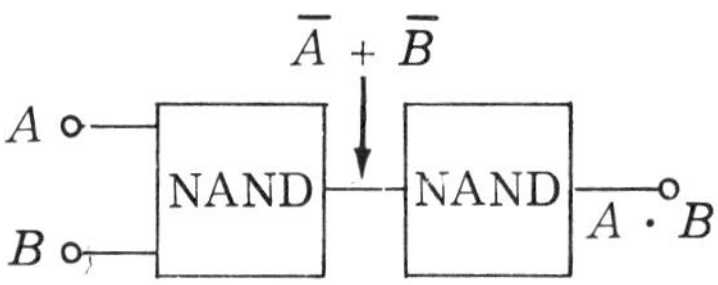

AND

$(\overline{\overline{A} + \overline{B}}) = A \cdot B$

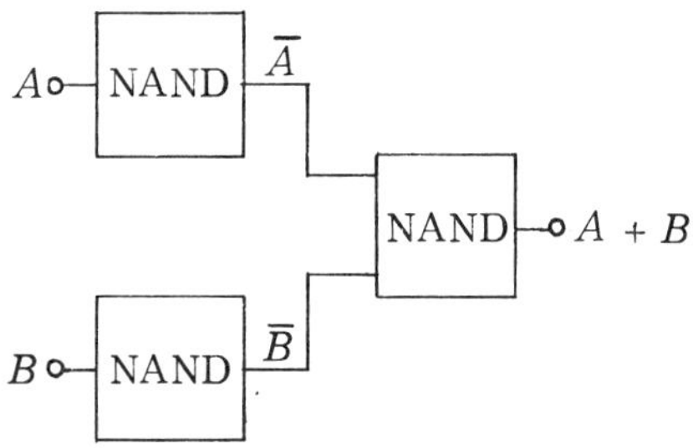

OR

$\overline{\overline{A}\,\overline{B}} = A + B$

If we wish to design circuits with NAND blocks, we could design the circuit with AND, OR, and INVERTER blocks and replace them by their NAND equivalents. Better circuits can usually be obtained by working backwards from the output block. For example, if we wanted to implement $A\overline{B} + CD$, the inputs to the output block would have to be $(\overline{A\overline{B}})$ and $(\overline{CD})$.

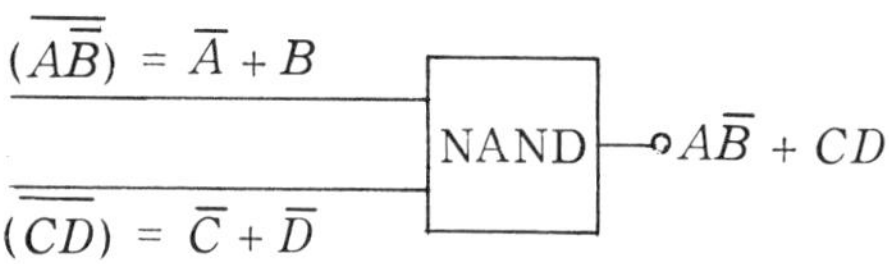

Completing the circuit we get:

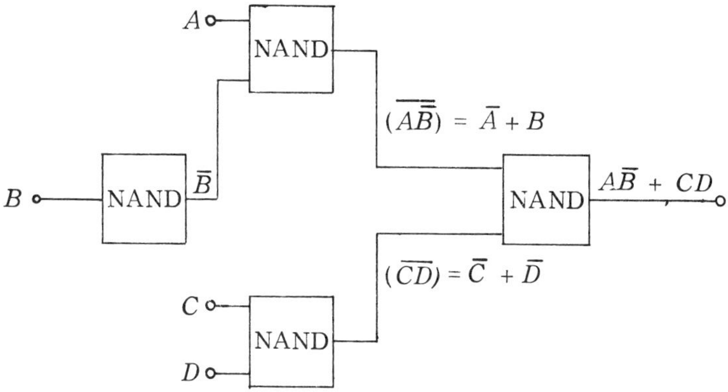

Some typical functionally complete sets are listed below. The sets marked with an asterisk (*) assume that the constants 1 and 0 are available as inputs.

AND-OR-INVERT	*MAJORITY-INVERT
AND-INVERT	NAND
OR-INVERT	NOR
*EXCLUSIVE OR-OR	*MINORITY
*EXCLUSIVE OR-AND	*IMPLICATION

PROBLEMS

1. Design a circuit, using AND, OR, and INVERTER blocks, which has two outputs specified as follows:

   $X_1\ (A, B, C, D) = \Sigma(1, 5, 9, 11)$
   $X_2\ (A, B, C, D) = \Sigma(1, 5, 8, 10, 11)$

   It can be assumed that inputs A and B are never equal to 1 at the same time.

2. Design a circuit, using only AND and INVERTER blocks, which satisfies the following specification (assume uninverted inputs A, B, C, and D only):

   The output is 0 when B is 1 and C is 1, or when A is 0 and D is 0. The output is 1 at all other times.

3. Design a circuit, using only the IMPLICATION block, which satisfies the following specification:

CD \ AB	00	01	11	10
00		1	∅	1
01		1		
11		1		
10	∅	∅	∅	1

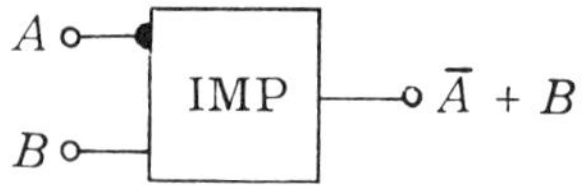

Implication block
(the semicircle denotes
the inverted input)

The implication block to be used must have two and only two inputs. (It can be assumed that fixed signal levels, representing the constants 0 and 1, may be used as inputs.)

4. Prove that the EXCLUSIVE OR-OR combination is functionally complete, assuming that fixed signal levels representing the constants 0 and 1 are available.

APPENDIXES

Appendix A

SYMBOLS

Boolean algebra is useful not only in logic design but also in many other fields. For example, mathematicians, logicians, and even lawyers are using it in their fields. It is therefore understandable that the different disciplines use different symbols. But, although the symbols are different, the same theorems and postulates are used. It is the purpose of this appendix to show other symbols found in the literature.

Constants

In this book:	0	1	
Other literature:	False	True	
	F	T	
	−	+	(Usually refers to voltages on wires)

Operations

Operation	Symbol	Example
AND (conjunction):		
In this book..........	no symbol or ·	XY, $X \cdot Y$
In other literature....	&, $\cap$, or $\wedge$	$X \& Y$, $X \cap Y$, $X \wedge Y$
OR (disjunction):		
In this book..........	+	$X + Y$
In other literature....	$\cup$, $\vee$	$X \cup Y$, $X \vee Y$
INVERSION (negation, complementation):		
In this book..........	bar over variable	$\overline{X}, \overline{\overline{X}Y} = \overline{(\overline{X})Y}$
In other literature....	’ to right of variable	X', $(X'Y)'$
	~ to left of variable	$\sim X$, $\sim(X \cdot Y)$

Blocks

A frequently used set of logic block symbols are:

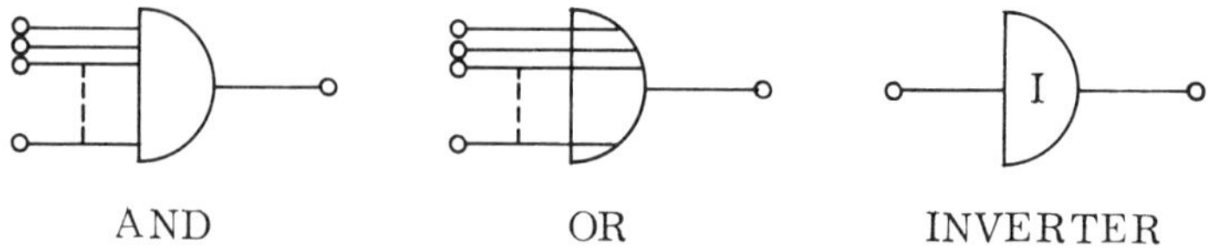

Note that the difference between the OR and the AND is the fact that in the OR block all inputs are drawn through to the other side.

Appendix B

THE BINARY NUMBER SYSTEM

One of the most important jobs of a digital computer is to perform arithmetic operations on numbers. To do this, numbers must be represented by signals which have only two states. One way of doing this is to have each decimal digit represented by four lines. Since the four lines may have any one of 16 values, and only 10 values are needed to indicate a decimal digit, 6 of the 16 values would not be allowed.

Another way to represent numbers in a computer is to make use of a different number system. The choice of the decimal system is generally attributed to the simple fact that man has ten fingers, and these fingers were used as the ten "digits" when counting. Since computers do not count on fingers, it is reasonable to assume that the decimal system may not be the best choice. In fact, the binary number system is much better suited for computers whose signals are limited to two values.

Let us first study the decimal number system, because there are many simularities between it and the binary number system. The decimal system; which derives its name from the Latin *decem* [ten; uses 10 different symbols (0 to 9]. Thus, when counting in this number system, we start with 0 and count up to 9. When we get past 9, we run out of new symbols and require a second digit. To write the number which is 1 higher than 9, we write 10 and mean that 10 equals $1 \times R + 0$, where R is the radix (or base) of the number system, which is one greater than the largest symbol. With two digits, we can count up to 99. The next higher number is 100, which is $1 \times R^2 + 0 \times R + 0$ or $1 \times 10^2 + 0 \times 10 + 0$. In general, it can be deduced that the first digit is multiplied by R^0 (which is equal to 1) the second is multiplied by R^1, the third is multiplied by R^2, and the nth is multiplied by R^{n-1}. The value of a number can be found by first multiplying each symbol by the appropriate multiplying factor (power of R) and then adding up all these products. From experience we know that omitting the leading zeros will not change the result. Only those terms whose symbol is something other than zero add to the value of the number. Thus all terms with the symbol 0 will be zero and can be omitted from the summation process. This property, that leading zeros will not change the value of the number, is true for the binary number system for the same reason, and it will not be discussed further.

Before going into the binary system, a new notation will be introduced to designate which number system we are using. The radix of the number system will accompany the number; for example, $6_{(10)}$ is "six" in the decimal system; $10_{(2)}$ is a number "10" in the binary number system. It should be noted that the radix is always written in the decimal number system.

The binary system, which derives its name from the Latin prefix bi, meaning two, is similar to the decimal system, except that it has only two symbols (0 and 1). To count, we start at 0 and go to 1. At this point we run out of symbols and must resort to a second digit. The radix of the system is 2 (one higher than the highest symbol which is 1). A decimal 2 in the binary number system is written as 10. (This is read one zero; do *not* read ten). Working with the formula $10_{(2)} = 1 \times R + 0$, where $R = 2_{(2)}$ 10 in the binary system is $1 \times 2 + 0$ which is $2_{(10)}$. Continuing the count, $3_{(10)} = 11_{(2)}$. At this point we run out of symbols again and require a third digit. Following the pattern, $4_{(10)} = 100_{(2)}$. Now the counting continues 101, 110, 111. Again we need another digit. The next numbers are 1000, 1001, 1010, etc.

Numbers given in the binary system can easily be translated into the decimal system. We know that the right-most binary digit is multiplied by $2^0(1)$, the second digit from the right is multiplied by 2^1 (which is 2), the next digit by 2^2 (which is 4), and the nth digit is multiplied by 2^{n-1}. For example, translating the binary number 1100111 into a decimal number gives:

1	1	0	0	1	1	1	Binary number
R^6	R^5	R^4	R^3	R^2	R^1	R^0	Multiplier for each digit
2^6	2^5	2^4	2^3	2^2	2^1	2^0	
64	32	16	8	4	2	1	
1×64	1×32	0×16	0×8	1×4	1×2	1×1	Product of multiplier times values of the digit position
64	32	0	0	4	2	1	

The decimal number is the sum of these numbers:

$$64_{(10)} + 32_{(10)} + 4_{(10)} + 2_{(10)} + 1_{(10)} = 103_{(10)}$$

Thus

$$1100111_{(2)} = 103_{(10)}$$

Translating decimal numbers into binary numbers can be accomplished by the following method.

To start the method, write the decimal number and draw a vertical line to its right. Divide the decimal number by 2 and place the remainder of the division on the right of the line. For example, 29 divided by 2 is 14, remainder 1. Continue this process until you end up with the number 0 on the left-hand side of the line. The binary number corresponding to decimal 29 is found by reading the binary digits from the bottom up: $29_{(10)} = 11101_{(2)}$.

First step

29	1
14	

Full problem

29	1
14	0
7	1
3	1
1	1
0	

Here are some equivalences you can check as practice problems:

Decimal numbers	Corresponding binary numbers
5	101
12	1100
14	1110
23	10111
93	1011101
303	100101111
3114	110000101010
4095	111111111111
4096	1000000000000

This appendix has dealt only with the translation of decimal integers into binary integers and vice versa, because the book does not require any further knowledge of the binary number system. However, it is possible to write fractions and to add, subtract, multiply, and divide in the binary number system. Since all other operations can be built up of these four operations, anything that can be done in the decimal number system can be done in the binary system.

A disadvantage of the binary system is that a binary number has more digits than an equivalent number in the decimal system (about 3.3 times as many); it is easier, however, for a computer to represent binary numbers, and it is also easier to design circuits which perform binary arithmetic operations. As an aid, the first 63 binary numbers, and their decimal equivalents, are listed in the Decimal to Binary Conversion Table at the back of the book.

Appendix C

DIODE CIRCUITS

Definitions

1. Current flow convention: Current flows from the more positive to the more negative voltage. (This is in the direction opposite to the flow of electrons.) The more positive level will be given the symbol 1, and the more negative level will be given the symbol 0.

2. Symbols:

Resistor

Diode

3. Diode characteristics: In an ideal diode (which we will assume), current can flow with zero resistance in the direction of the arrow—but cannot flow (i.e., meets infinite resistance) in the opposite direction.

1Q In the following circuit, we will apply the more positive voltage level at terminal a and the more negative voltage level at terminal b; we want to find the resulting voltage level at terminal x. We know that current flows from 1 to 0, and we know that the diode has zero resistance to current flowing in the direction of the arrow. It is also true that when there is zero resistance between two points, the voltages at those two points must be the same. Using this information, determine the voltage level (1 or 0) at terminal x.

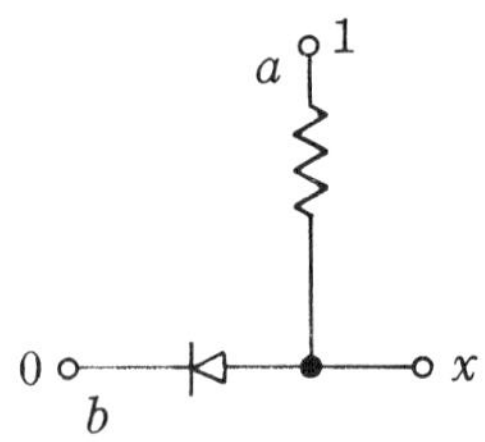

1A 0 (the more negative level)

Current (the amount of which is determined by the value of the resistor) tends to flow from terminal *a* to terminal *b*. In this direction, the diode has zero resistance. Since there is zero resistance between terminals *b* and *x*, the voltage must be the same at both terminals; the the voltage at terminal *x* is 0.

2Q Whenever terminals *a* and *b* are both at the same voltage level, then terminal *x* must also be at that voltage level. The voltage level at terminal *x* in the circuit shown is ______.

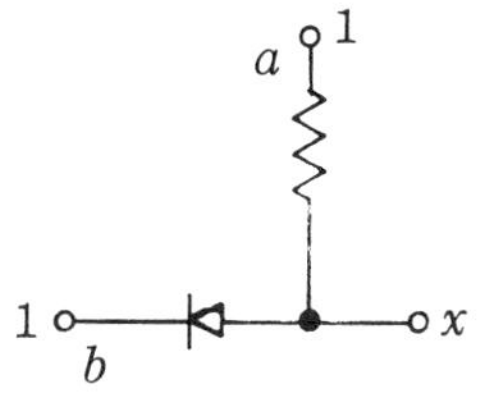

2A 1 (positive)

Since there is no voltage difference between terminals *a* and *b*, no current flows and all points in the circuit are at the same voltage level.

3Q Let us now study the two-diode circuit shown here. The voltage level at terminal *a* of this circuit will always be at the 1 level, but the levels at terminals *b* and *c* can be changed. We are interested in finding out what the level at terminal *x* will be when terminals *b* and *c* are set to particular levels. For example, when terminals *b* and *c* are both at the 0 level, as shown, the level of *x* is at ______.

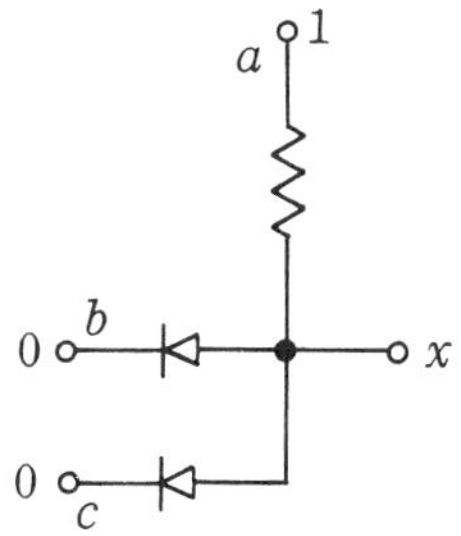

3A 0 (negative)

Current tends to flow from terminal *a* to terminals *b* and *c*. In this direction, both diodes have zero resistance; thus *x* is at the negative voltage level.

4Q The binary voltage level at terminal *x* in the circuit shown is ______.

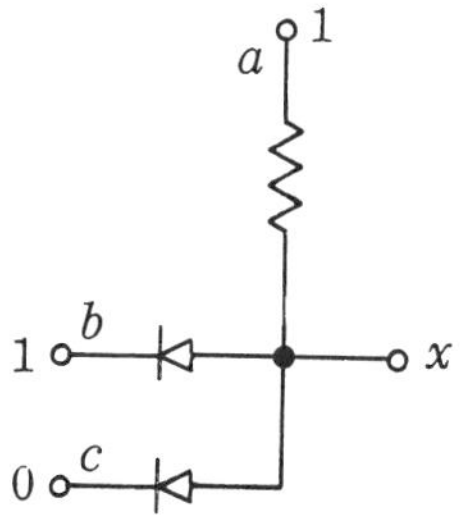

4A 0 (negative)

Current tends to flow from terminal *a* to terminal *c*, the direction of the diode arrow at *c*. Since the diode at *c* has zero resistance, the level at *x* must be the same as the level at *c*, and so the level at *x* is 0. With a 0 level at *x* and a 1 level at *b*, current would tend to flow in a direction opposite to that of the arrow of the upper diode; in this direction, however, the diode has infinite resistance, so the voltage at *b* is effectively disconnected and cannot affect the voltage at *x*.

5Q If the levels at terminals *b* and *c* are interchanged, as shown, the level at terminal *x* is ______.

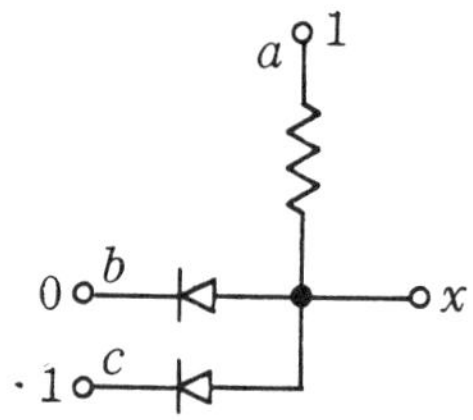

5A 0 (negative)

The reasoning used in the frame 4A can be repeated here if terminals *b* and *c* are interchanged.

6Q If terminals *b* and *c* are both at a 1 level, the level at terminal *x* is ______.

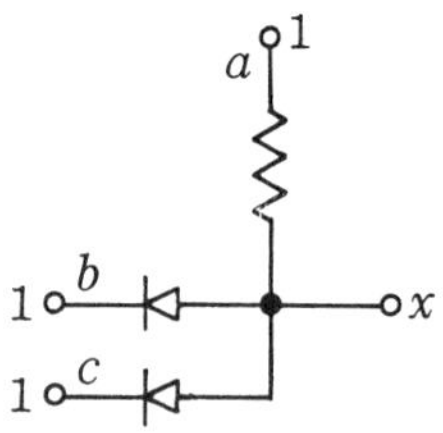

6A 1 (positive)

Since terminals *a*, *b*, and *c* are all at the 1 level, no current flows, and *x* must also be at the 1 level.

7Q We have now determined the level at terminal *x* for every allowable combination of levels at terminals *b* and *c*. The results are as follows: *x* is 0 when (1) *b* is 0 and *c* is 0, (2) *b* is 1 and *c* is 0, (3) *b* is 0 and *c* is 1; *x* is 1 when *b* is 1 and *c* is 1. Can these characteriestics be completely specified merely by saying that *x* is 1 only when both *b* and *c* are 1?

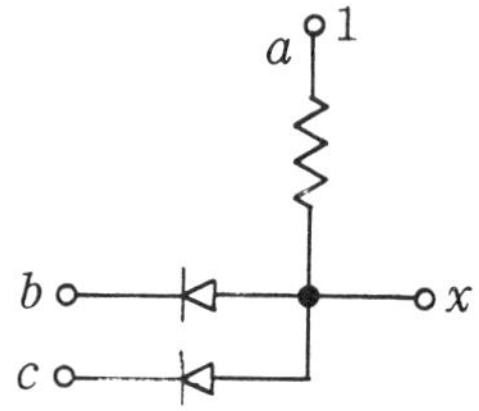

7A Yes

8Q Although the circuit has been shown with two diodes, it actually can have any number of diodes. We shall now determine the input characteristics of the circuit below for the general case where the number of diodes is not specified. If one or more of the terminals *b* through *m* are 0, what is the level at terminal *x*?

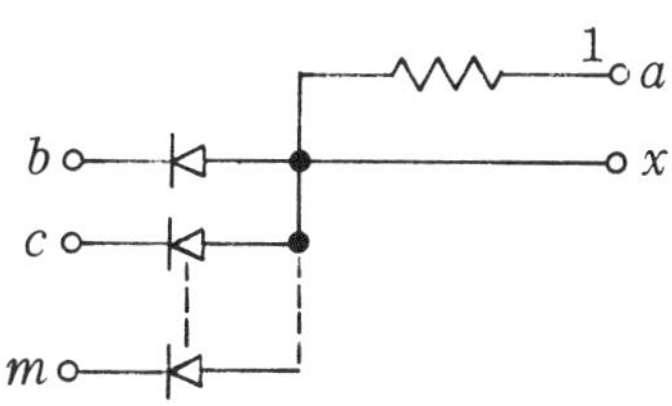

8A 0 (negative)

Current tends to flow from terminal a, which is always positive, to any terminal which is at a negative level. This current flows in the direction of the diode arrow, meaning that there is zero resistance between the negative terminal and x; thus x will be negative. The remaining diodes will not come into play since the current tends to flow against the arrow.

9Q What is the level at x if all terminals b through m are positive?

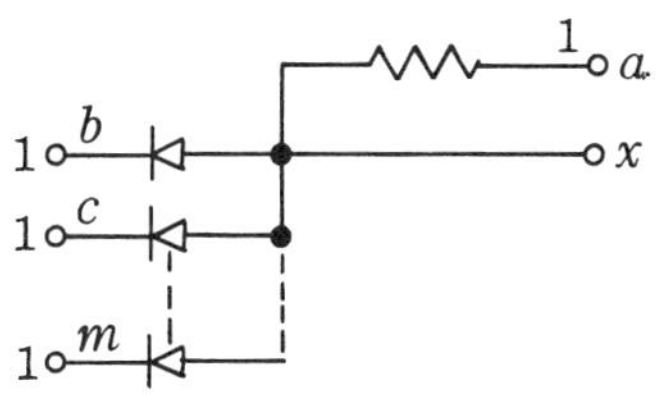

9A 1 (positive)

Since all terminals are positive, no current flows and all points within the circuit must be positive.

10Q In the general case, then, terminal x is at the 1 level only when all terminals b through m are 1. Another way of stating this is to say that x is at the 0 level when ____________________.

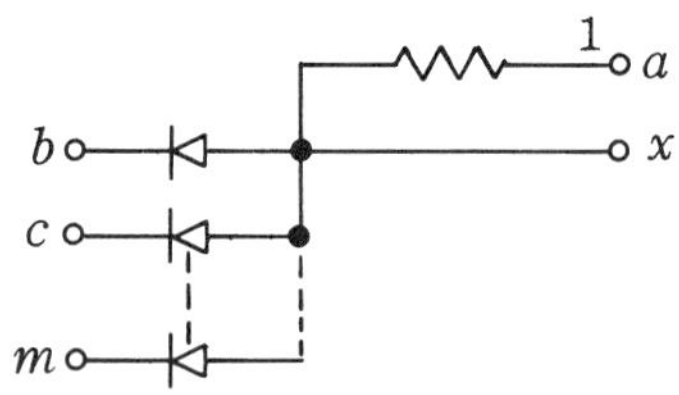

10A One or more terminals *b* through *m* are 0

11Q In Chap. 1, it was mentioned that there are different types of logic blocks. One of these blocks is defined by the fact that its output is 1 if and only if all inputs are 1. This logic block is called an AND block because its output is 1 only if the first input *and* the second input *and* the third input, etc., are 1. Can the diode circuit just analyzed be used for this logic block?

11A Yes

If we define terminals *b* through *m* of the diode circuit as inputs and terminal *x* as the output, then the characteristics of this diode circuit are identical with the required characteristics of the logic block, and the circuit may be used as an AND block.

12 The symbol for the AND block is shown, along with the diode circuit drawn in such a way that the terminals of the diode circuit can be related to the logic block inputs and outputs.

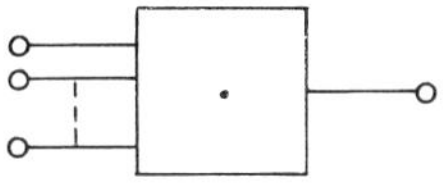

Logic block

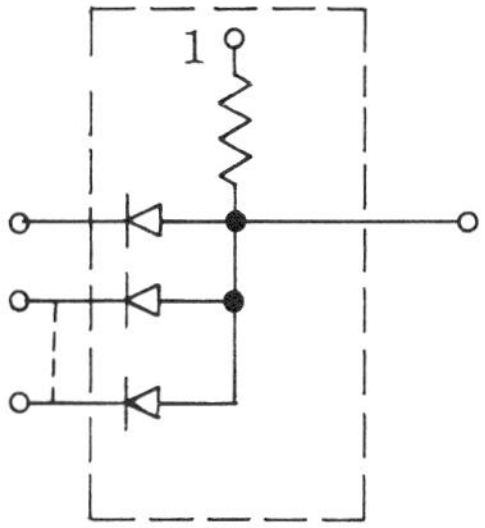

Diode circuit

Note that one point is labeled 1. This point in the diode circuit is connected directly to a power supply. Since it is not an input, this point does not appear in the logic block representation.

---

13 Let us now turn our attention to the second of the three circuits to be studied.

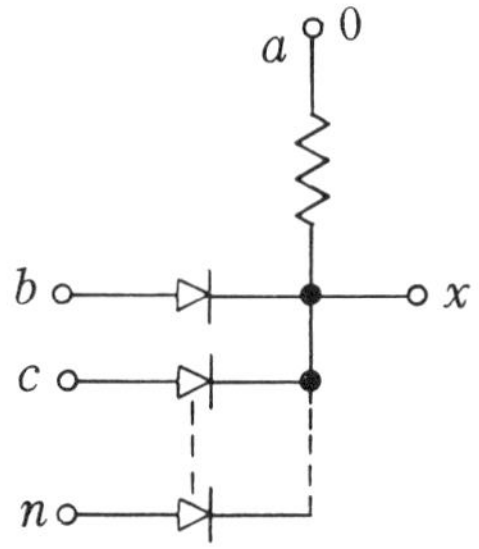

There are two differences between this circuit and the previous one:

1. Terminal a is permanently connected to a 0 voltage.

2. The diodes are connected so that the arrows point to terminal x.

14Q What is the level at terminal x of the circuit of frame 13 when terminals b through n are all at the 1 level?

14A 1 (positive)

When terminals b through n are positive, current tends to flow from these terminals to terminal a. Since this is the direction of the diode arrow, there is zero resistance between terminals b through n and terminal x; thus terminal x is at the 1 level.

15Q When only one terminal (for example, b) is at the 1 level, the level at terminal x is______.

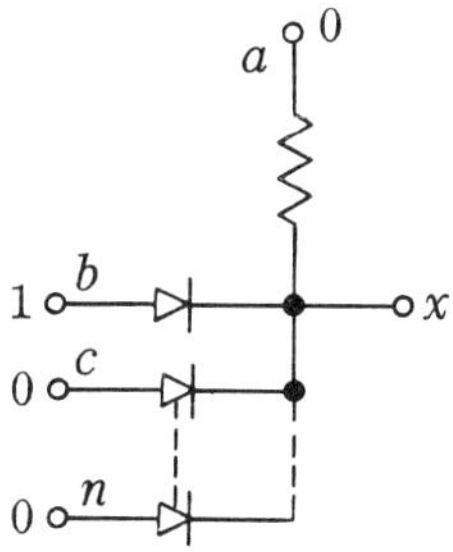

15A 1 (positive)

Any diode having zero resistance is sufficient to make x positive. All other diodes tend to conduct in the direction opposite the direction of the arrow, and since they have infinite resistance in this direction, they do not affect the circuit.

16Q In the diagram, what is the level at terminal x when terminals b through n are all 0?

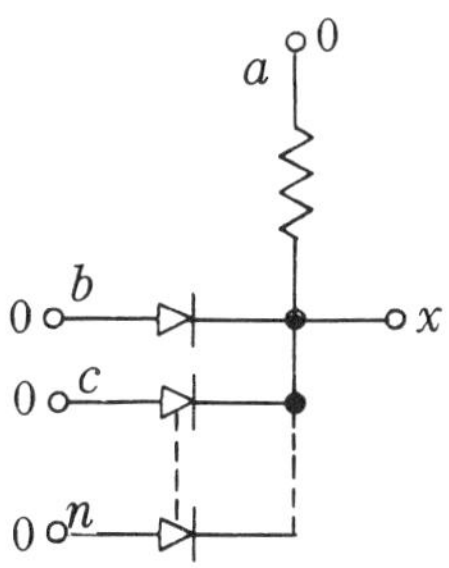

16A 0 (negative)

If all terminals are negative, no current flows, and x must be negative.

17Q In summary, it can be said that terminal x will be at the 1 level when ______________________.

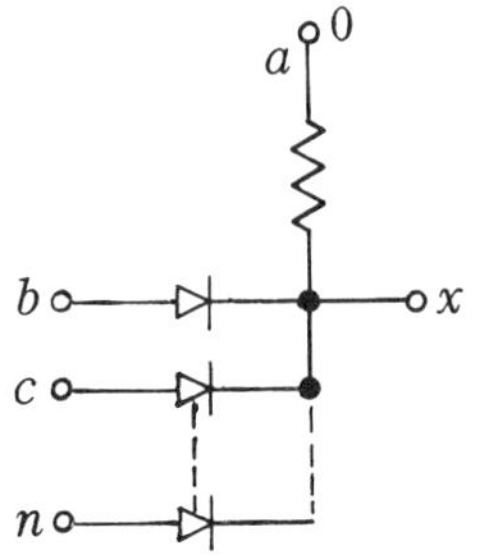

17A One or more of the terminals b through n are at the 1 level.

18Q Terminal x will be at the 0 level when ________________.

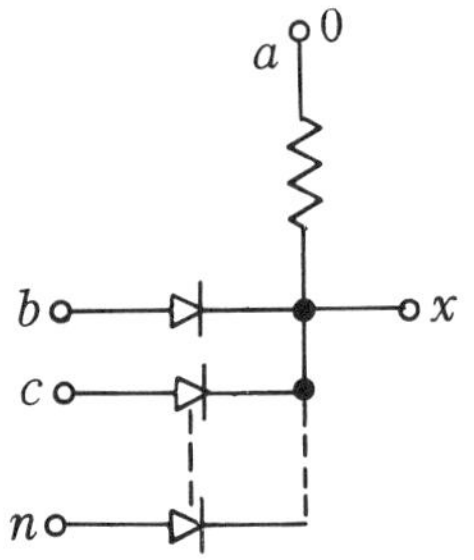

18A *all* terminals b through n are at the 0 level.

19 The logic block for which this circuit can be used is called the OR block, which is drawn on page 271.

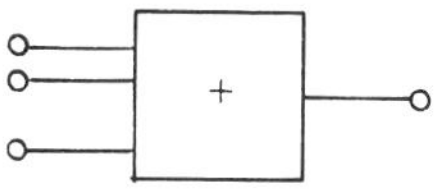

Logic block

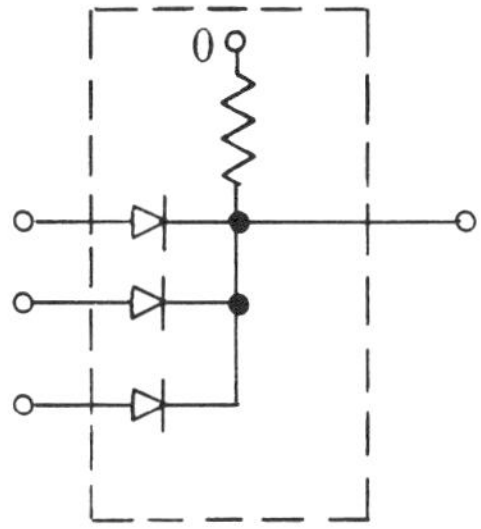

Diode circuit

This block is called an OR block because its output is 1 if the first *or* the second *or* the third input, etc., *or* any combination of these inputs, is 1.

---

20 The third circuit to be covered in this chapter has both a triode vacuum tube and a transistor version. Most recent computers make use of the transistor version, but the tube circuit is still important and will be included for those readers more familiar with tubes than with transistors. The symbols to be used for vacuum tubes and transistors are:

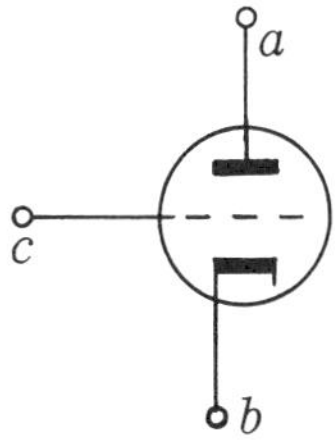

Vacuum tube

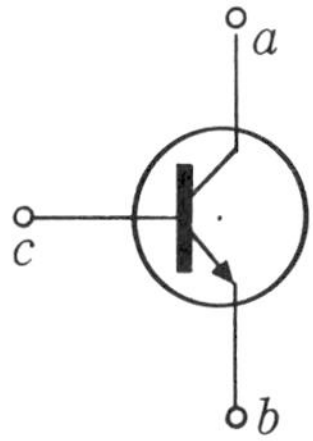

Transistor

Both devices act in a similar manner. The voltage at terminal *c* controls the flow of current from terminal *a* to terminal *b*. (The arrow at terminal *b* of the transistor again represents the direction of current flow.) No matter what voltages are applied, current can never flow between *c* and the other terminals. The basic circuits in which these devices are used are shown below. Although additional components (resistors and capacitors) are usually added to the circuits, they are not essential in this discussion.

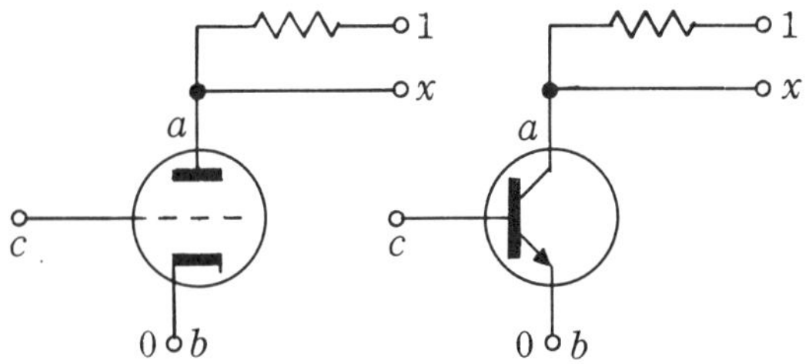

21Q In the ideal case, there is zero resistance between terminals *a* and *b* when terminal *c* is at the positive level.

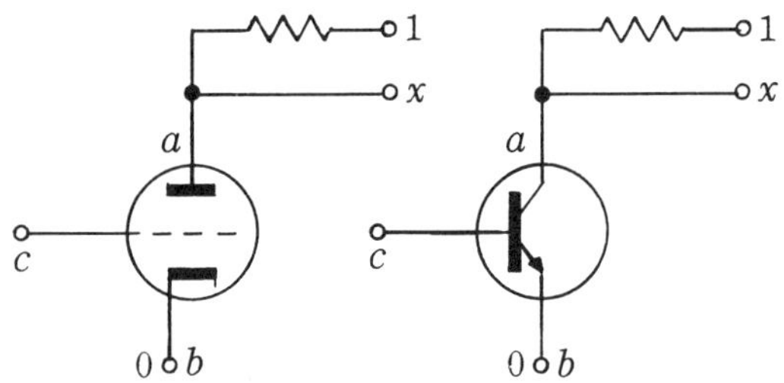

With zero resistance between terminals *a* and *b*, the level at *x* would be ______.

21A 0 (negative)

If there is zero resistance between two points, then the voltages at those two points must be the same.

22Q A 1 level at *c* causes zero resistance between *b* and *x*. Thus a 1 level at *c* results in a ______ level at *x*.

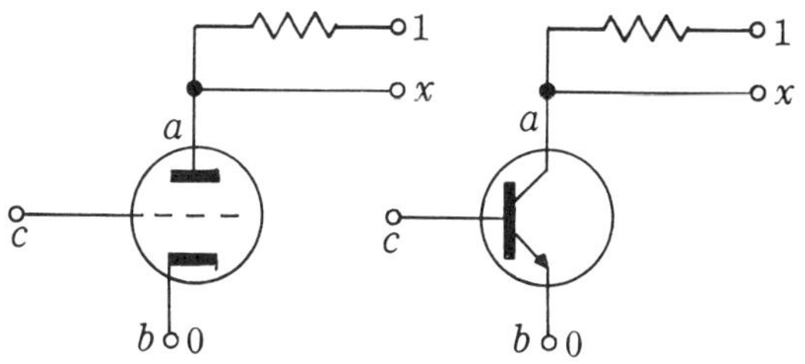

22A 0 (negative)

23Q

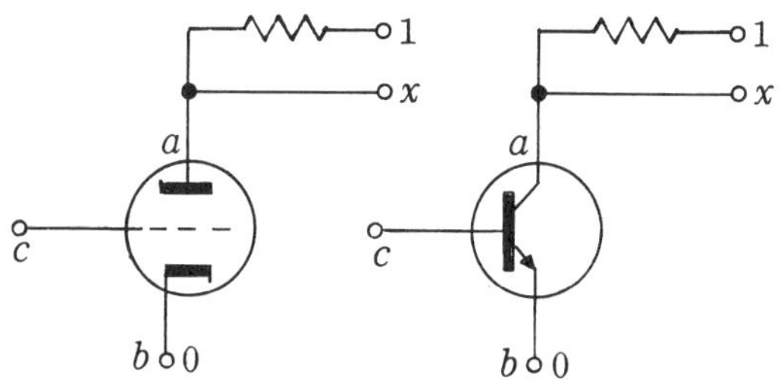

If terminal c is at the 0 level, the resistance between terminals a and b is, ideally, infinite. As a result, no current flows in the circuit. If no current flows through the resistor, then the voltage at both ends of the resistor must be the same. This is no different than saying that if the voltage at both ends of a resistor is the same, then no current flows through it. When terminal c is at the 0 level, the level at x is______.

23A 1 (positive)

When c is 0, the resistance between terminals a and b is infinite, and no current flows in the resistor. This means that both ends of the resistor must have the same voltage.

24 The operation of this circuit can be summarized as follows:

A 1 level at c results in a 0 level at x.
A 0 level at c results in a 1 level at x.

The output of this circuit is the opposite, or inverse, of the input; thus the circuit is called an INVERTER. The logic block for the INVERTER is a square with one input and one output:

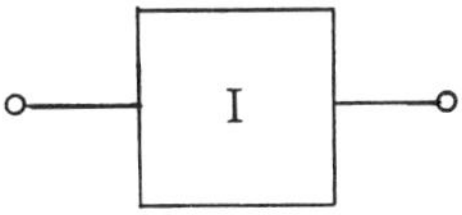

25 The following table is a summary of the contents of this appendix.

Circuit example	Logic block	Name	Characteristics
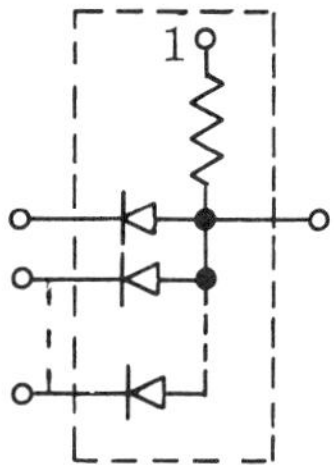	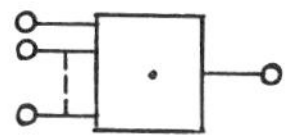	AND	Output is 1 if *all* inputs are 1; output is 0 if any input is 0
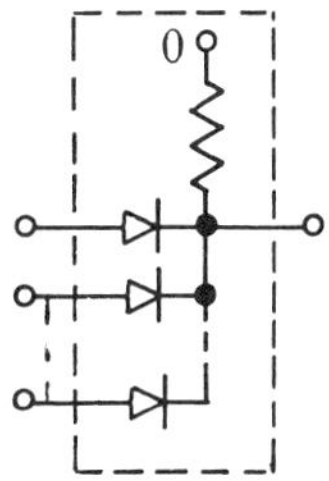	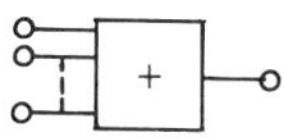	OR	Output is 1 if one or more inputs are 1; output is 0 if all inputs are 0
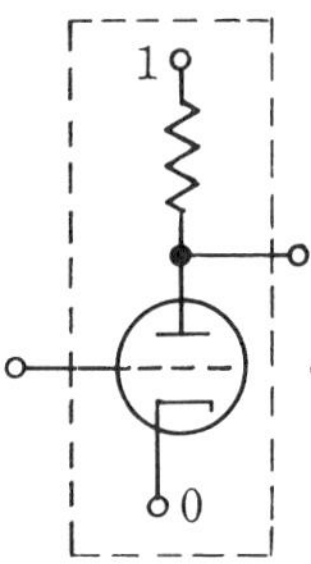 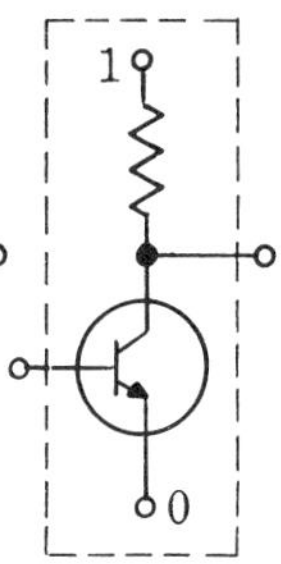	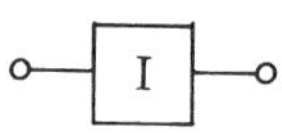	INVERTER	Output is 1 if input is 0; output is 0 if input is 1

Appendix D

RELAY LOGIC

Relays have two main parts: the relay *coil* and the relay *contact* (or set of contacts).

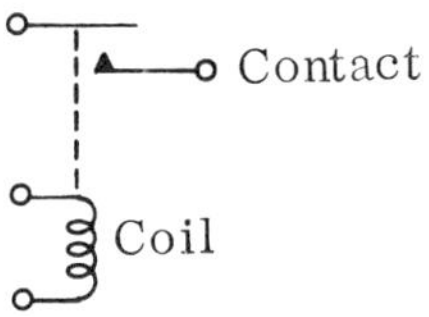

The parts of the contact are generally called ***springs*** because they are made of a flexible springlike metal.

When current passes through the relay coil, the magnetic field created by the current pulls the movable upper contact spring down against the lower fixed spring. In this position there is a short circuit (no resistance) between the two terminals of the contact. When current stops flowing through the coil, the upper spring will return to the position shown. In this position, there is an open circuit (infinite resistance) between the two terminals of the contact.

The contact just described is called a *normally open contact*. The "normal" condition occurs when *no* current flows through the coil. A relay can also be built with a *normally closed contact*.

Normally open contact

Normally closed contact

When current is *not* flowing in the relay coil (the "normal" condition), the normally closed contact forms a closed circuit. When current *is* flowing in the relay coil, the movable lower spring is pulled away from the fixed upper spring to form an open circuit.

Another type of relay contact is called the *transfer contact.*

Transfer contact

When current is not flowing in the relay coil, the movable center spring is in contact with the fixed upper spring. When current flows through the coil, the movable center spring is pulled down by the magnetic field and makes contact with the lower fixed spring.

Contacts are always drawn in their normal position, that is, the position they are in with no current flowing in the relay coil.

It is possible for a single relay coil to control more than one contact. In the following diagram, a relay coil controls three contacts: a normally open, a normally closed, and a transfer contact.

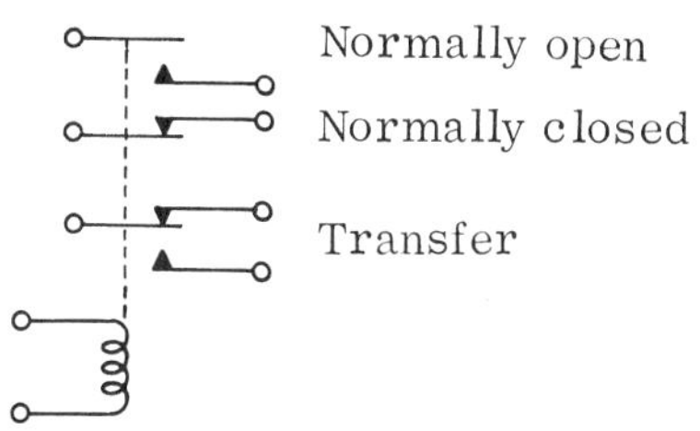

Relay Circuits

The importance of relays lies in the fact that the contacts of one relay can be used to control the flow of current through the coil of another relay. In general, current through the coil of one relay can be controlled by the contacts of any number of relays. In the following example, the contacts on the two relays *A* and *B* control the current through the coil of a third relay, *C*.

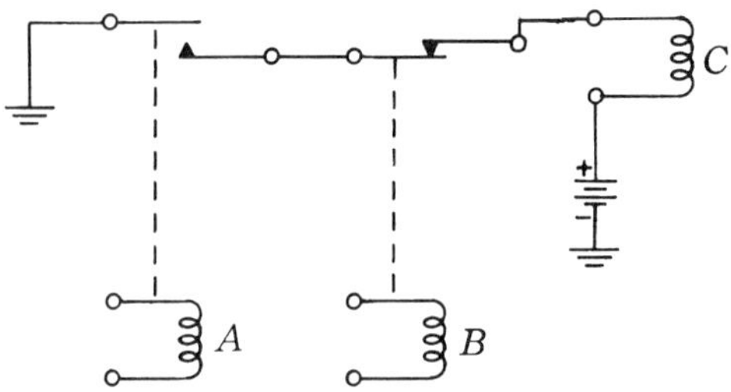

In this particular circuit, battery current flows through the coil of C only if the contacts on relays A and B supply a path for that current. If current flows through the coil of relay A and no current flows through the coil of relay B, then the contacts on both of these relays will form short circuits, and a path for current through coil C will be formed. The contacts of relay C (which are not shown) can be used to control current through the coils of other relays. The contacts which control the flow of current through a relay coil form the *contact network* for that relay.

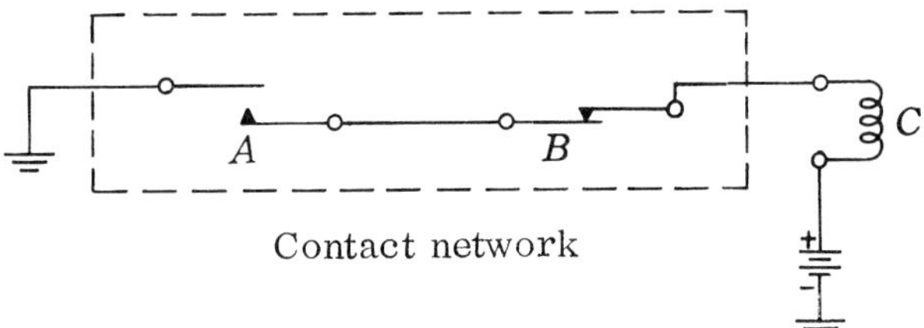

Contact network

Depending on the states of all the contacts, there is either a closed circuit or an open circuit between the two terminals of the contact network. A closed circuit allows current to flow through the coil, while an open circuit prevents the flow of current. The contacts in a contact network may be identified by drawing a dotted line between each contact and its coil, or by assigning a letter to each coil and writing that letter near each of its contacts.

Application of Boolean Algebra to Relay Logic

When applied to relay circuits, the constants, variables, and operations of Boolean algebra have a different meaning than they do when applied to networks of logic blocks. In relay circuits, the constants 0 and 1 represent an open circuit and a closed circuit, respectively. This may be applied to a single contact or to a complete contact network. If there is an open circuit between two terminals, the contact, or contact network, is in the 0 state. If there is a short circuit between two terminals, the contact, or contact network, is in the 1 state.

The constants may also be used to represent the state of current in a relay coil. The coil is in the 0 state if no current is flowing through it; it is in the 1 state if current is flowing through it. When a contact network controls the current through a relay coil, both the network and the coil have the same value. That is, a contact network value of 1 means a closed circuit, which allows current to flow through the coil, giving it a value of 1. A contact network value of 0 means an open circuit, which prevents the flow of current, giving the coil a value of 0.

A variable is used to describe the state of a relay coil. For example, when $A = 0$, relay A has no current flowing through its coil. When $A = 1$, the relay has current flowing through its coil. The variable that describes the state of a relay coil can also be used to describe the state of all normally open contacts on that coil. When $A = 1$, current is flowing in the coil, which means that the normally open contacts are closed and that they also have the value of 1. When $A = 0$, no current is flowing in the coil, which means that the normally open contacts are open and they also have the value of 0.

Since a normally closed contact will always be in the state opposite that of a normally open contact on the same relay, the closed contact can be described by the inverted form of the variable. For example, when $A = 1$, current is flowing through the coil and all normally closed contacts are open and have the value of 0. Since $\overline{A} = 0$ when $A = 1$, the state of the normally closed contact is correctly described by $\overline{A}$. The labeling of a coil and its contacts is shown on page 280.

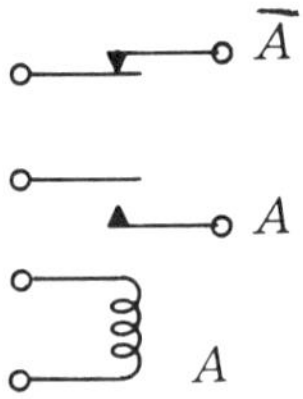

The Boolean operations of AND and OR are used to describe the *connections* between contacts. If two or more contacts are connected in series with each other, the AND operation is performed.

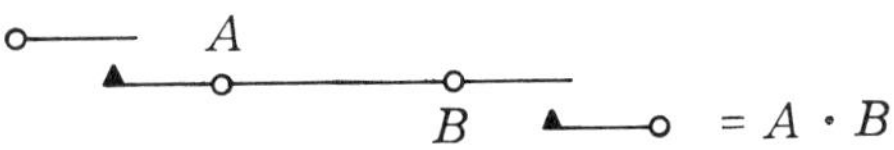

The circuit is shorted only if both contact A *and* contact B are shorted. Since a short circuit has the value of 1, this can be restated as: The circuit has the value of 1 only if A is 1 *and* B is 1. We can therefore describe the circuit by $A \cdot B$, since $A \cdot B$ is 1 only if A is 1 *and* B is 1. Describing the circuit by $A \cdot B$ is a way of saying that the circuit is shorted only when current is flowing in relay coil A *and* relay coil B.

If normally closed contacts are connected in series, the connection is still described by the AND operation.

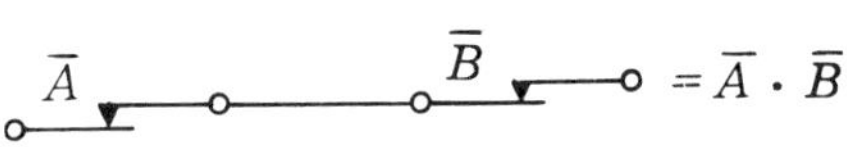

The circuit equals 1 only when $\bar{A}$ equals 1 *and* $\bar{B}$ equals 1. This means that the value of the circuit is 1 only when current is *not* flowing in relay coil A and current is *not* flowing in relay coil B.

If two or more contacts are connected in parallel, the OR operation is performed.

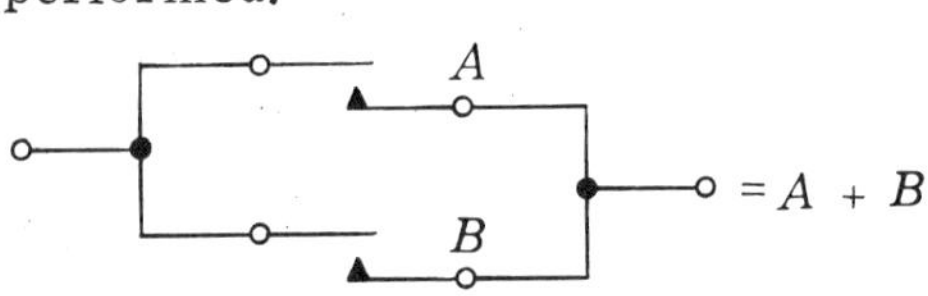

The circuit is shorted only if contact A is shorted *or* contact B is shorted. In other words, the circuit equals 1 only if A is 1 *or* B is 1. The circuit can therefore be described by $A + B$. A is 1 when current flows through coil A and B is 1 when current flows through coil B, so current through either coil will cause the value of the circuit to equal 1.

A contact network may consist of contacts connected in both series and in parallel.

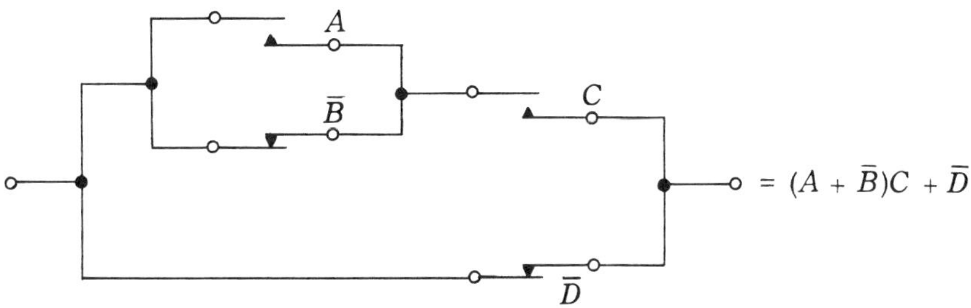

A relay coil must always have the same value as its contact network. When the contact network has the value of 1, current will flow through the relay coil, so the relay coil also has the value of 1. When the contact network has the value of 0, no current can flow through the coil, so it also has the value of 0. This relationship may be expressed by an equation, since both sides of an equation are also equal.

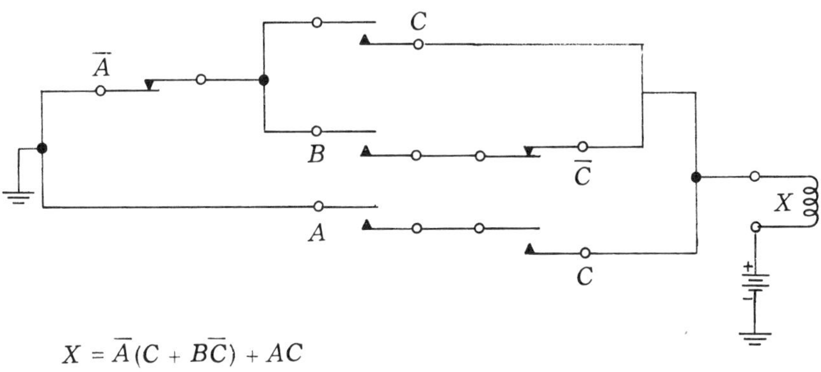

$$X = \bar{A}(C + B\bar{C}) + AC$$

Using Boolean theorems or other minimizing techniques, the equation can be simplified and a better circuit can be obtained.

$$
\begin{aligned}
X &= \overline{A}C + \overline{A}B\overline{C} + AC \\
&= (\overline{A} + A)C + \overline{A}B\overline{C} \\
&= C + \overline{A}B\overline{C} \\
&= C + \overline{A}B
\end{aligned}
$$

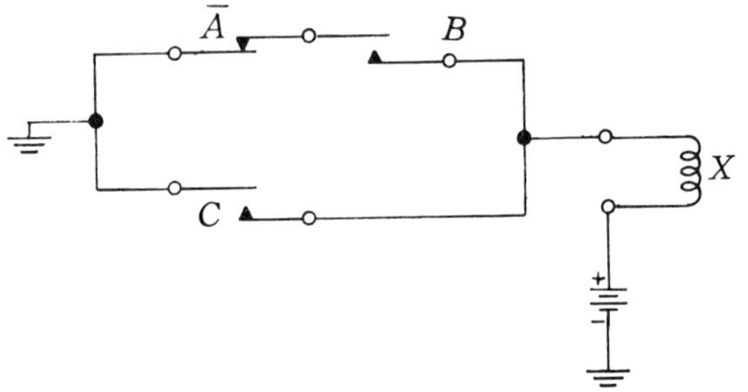

ANSWERS TO PROBLEMS

Chapter 2

1. $A = 1;\ B = 0;\ \overline{C} = 0$

2.

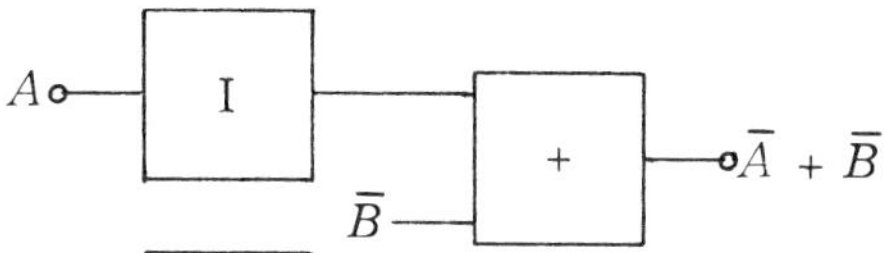

3.

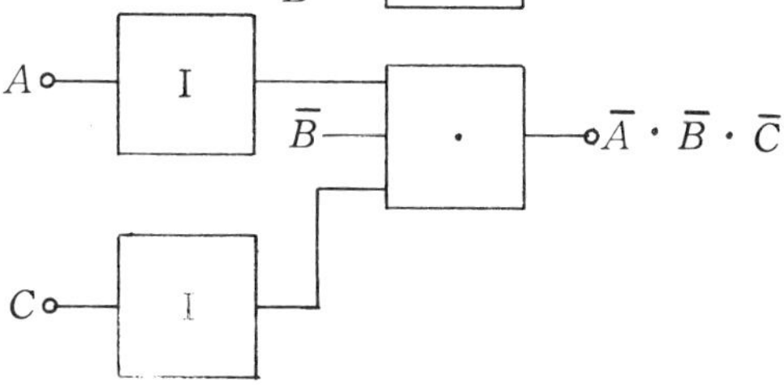

4.

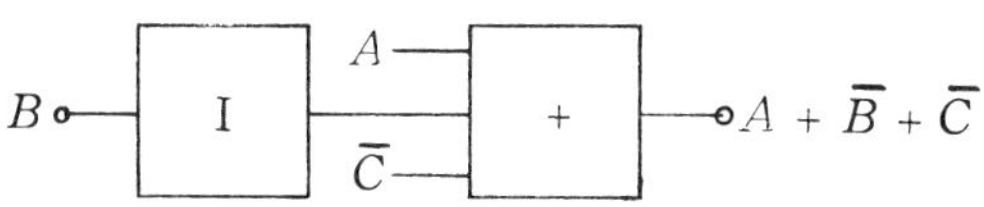

5. $A = 0;\ B = 0$ 6. $A = 1;\ B = 1$

Chapter 3

1.

A	B	$A + \overline{B}$
0	0	1
0	1	0
1	0	1
1	1	1

2. Four rows; A and $\overline{A}$ are different forms of the same variable.

3.

A	B	C	$\overline{C}$	$A \cdot B \cdot \overline{C}$
0	0	0	1	0
0	0	1	0	0
0	1	0	1	0
0	1	1	0	0
1	0	0	1	0
1	0	1	0	0
1	1	0	1	1
1	1	1	0	0

4.

A	B	C	$\overline{B}$	$\overline{C}$	$A+\overline{B}+C+\overline{C}$
0	0	0	1	1	1
0	0	1	1	0	1
0	1	0	0	1	1
0	1	1	0	0	1
1	0	0	1	1	1
1	0	1	1	0	1
1	1	0	0	1	1
1	1	1	0	0	1

(*a*)

A	B	C	$\overline{B}$	$\overline{C}$	$A \cdot \overline{B} \cdot C \cdot \overline{C}$
0	0	0	1	1	0
0	0	1	1	0	0
0	1	0	0	1	0
0	1	1	0	0	0
1	0	0	1	1	0
1	0	1	1	0	0
1	1	0	0	1	0
1	1	1	0	0	0

(*b*)

5. (*a*) $A \cdot (\overline{B} \cdot 0) = A \cdot 0 = 0$ since $\overline{B} \cdot 0 = 0$
(*b*) $A + A + \overline{A} + 1 = A + \overline{A} + 1$ since $A + A = A$
$= 1 + 1$ since $A + \overline{A} = 1$
$= 1$

6. (*a*) B; (*b*) $\overline{A}$

Chapter 4

1.

A	B	$(A + B)$	$(A + \overline{B})$	$(A + B)(A + \overline{B})$
0	0	0	1	0
0	1	1	0	0
1	0	1	1	1
1	1	1	1	1

2. $A + AB = A$. AB is replaced by 0, giving $A + 0 = A$. This can be done because in the case where AB is *not* 0, that is, when $AB = 1$, then $A = 1$ and the value of AB does not matter anyway $(1 + AB = 1)$.

3. (*a*) 1; (*b*) 0

4. $A + \overline{B}C$

5.

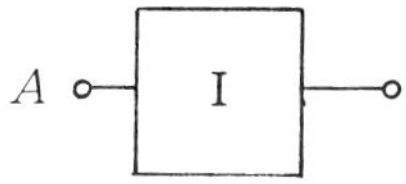

6.

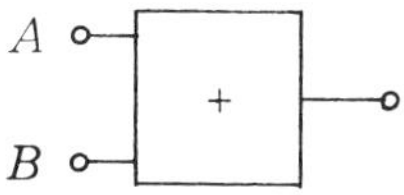

7.

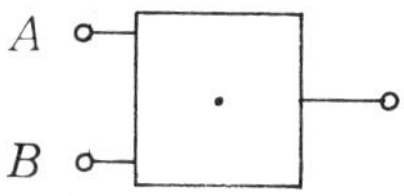

8. (a) 0

(b) $B(\overline{C + D})$

(c) $A + B + \overline{C}$

Chapter 5

1. (a) $\overline{A}\,\overline{B}(C + \overline{D})$; (b) $AB\overline{C}$; (c) $AB + \overline{C} + \overline{D}$; (d) $\overline{A}(\overline{B} + CD)$; (e) $[\overline{A} + (\overline{B} + \overline{C})D][\overline{A} + (B + C)\overline{D}]$

2. (a)

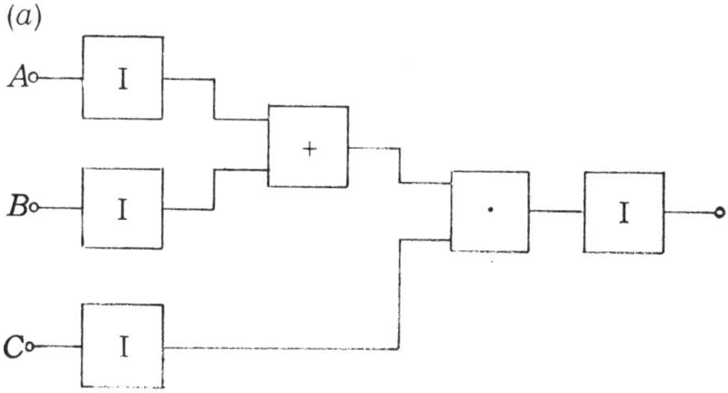

$AB + C = \overline{(\overline{A} + \overline{B})\overline{C}}$

(b)

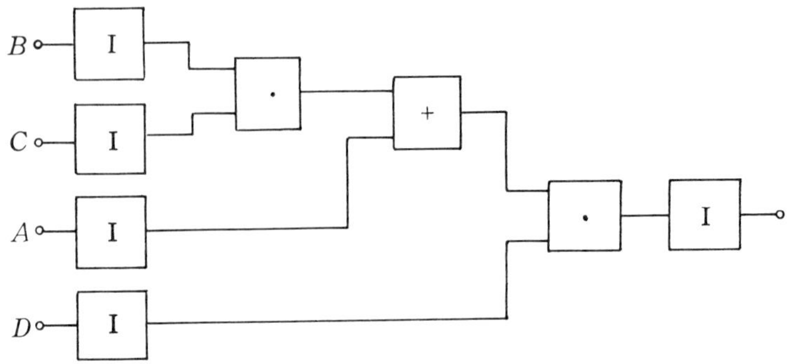

$$(B + C)A + D = \overline{(\overline{\bar{B}\bar{C}} + \bar{A})\bar{D}}$$

3. $W = \overline{\overline{(A + B)\bar{\bar{C}}}} = (A + B)\bar{C}$

4. $X = (A + \bar{B})(\bar{A} + B)\bar{C} = (AB + \bar{A}\bar{B})\bar{C}$

5.

A	B	C	ABC	$\overline{ABC}$	$\bar{A} + \bar{B} + \bar{C}$
0	0	0	0	1	1
0	0	1	0	1	1
0	1	0	0	1	1
0	1	1	0	1	1
1	0	0	0	1	1
1	0	1	0	1	1
1	1	0	0	1	1
1	1	1	1	0	0

The last two columns are equal therefore $\overline{ABC} = \bar{A} + \bar{B} + \bar{C}$

6. (a) $U = B$; (b) $V = C$

7. (a) $\bar{A}\bar{B} + \bar{A}\bar{C} = \bar{A}(\bar{B} + \bar{C})$; (b) $\bar{A}\bar{B} + \bar{B}\bar{C} + A\bar{C} = \bar{A}\bar{B} + A\bar{C}$

8. (a)

A	B	C	Before inversion		After inversion	
			Left side	Right side	Left side	Right side
0	0	0	0	0	1	1
0	0	1	0	0	1	1
0	1	0	0	0	1	1
0	1	1	1	1	0	0
1	0	0	1	1	0	0
1	0	1	1	1	0	0
1	1	0	1	1	0	0
1	1	1	1	1	0	0

(*b*)

A	B	C	Before inversion		After inversion	
			Left side	Right side	Left side	Right side
0	0	0	0	0	1	1
0	0	1	0	0	1	1
0	1	0	1	1	0	0
0	1	1	1	1	0	0
1	0	0	0	0	1	1
1	0	1	1	1	0	0
1	1	0	0	0	1	1
1	1	1	1	1	0	0

Chapter 6

1. $\underbrace{\text{Type out on the console typewriter}}_{Z}$ when (iff) $\underbrace{\text{the computer has completed the problem,}}_{A}$ $\underbrace{\text{there is a programming error,}}_{B}$ or $\underbrace{\text{an illegal operation is specified.}}_{C}$

$$Z = A + B + C$$

2. 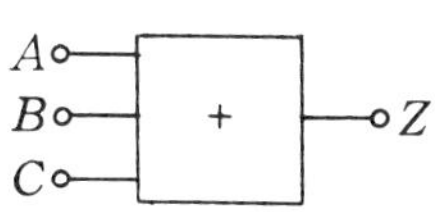

3. Variable Z: Mr. Jones will go to the movies
Variable A: he gets a baby sitter
Variable B: it rains
Variable C: it is a weekend
Variable D: a Western is playing

Combining the sentences into one:
Mr. Jones will go to the movies iff he can get a baby sitter, it does not rain, it is not a weekend, and a Western is playing.

Another version is:
Mr. Jones will go to the movies iff he can get a baby sitter, and a Western is playing, but not if it rains, or on weekends.

$$Z = AD(\overline{B + C}) \text{ or } A\bar{B}\bar{C}D$$

4. $\underbrace{\text{Store the content of the accumulator}}_{Z}$ when (iff) $\underbrace{\text{a STORE instruction is given,}}_{A}$ $\underbrace{\text{at the end of an operation when no error occurred,}}_{B}$ or when $\underbrace{\text{the console typewriter is used.}}_{C}$

$$Z = A + B + C$$

or

$\underbrace{\text{Store the content of the accumulator}}_{Z}$ when (iff) $\underbrace{\text{a STORE instruction is given}}_{A}$ $\underbrace{\text{at the end of an operation}}_{B}$ when $\underbrace{\text{no error occurred,}}_{C}$ or $\underbrace{\text{when the console typewriter is used.}}_{D}$

$$Z = A + BC + D$$

5.

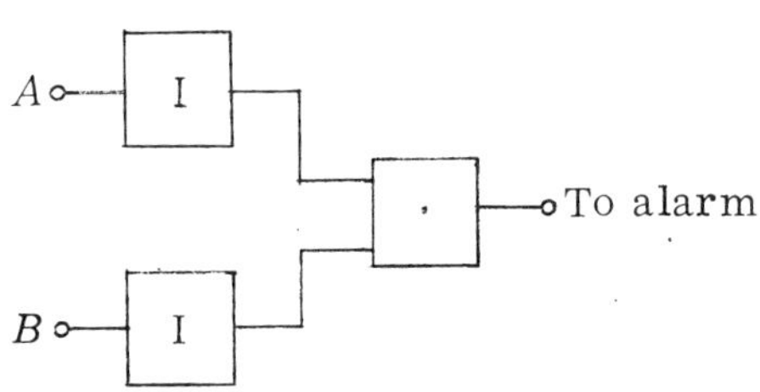

or

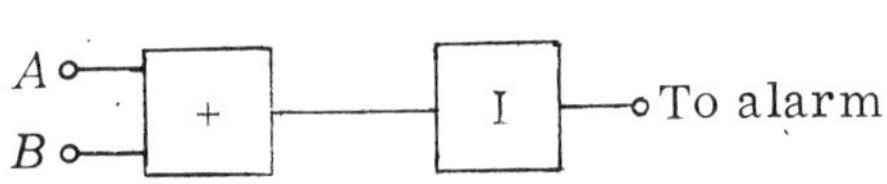

6. One signal is AB. Second signal is $A\overline{B} + \overline{A}B$.

$$A + B = \underbrace{AB}_{M} + \underbrace{(A\overline{B} + \overline{A}B)}_{N} = M + N$$

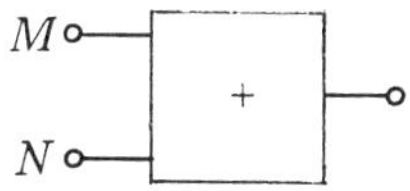

7. (*a*) $Z = A + B + C + D$; (*b*) $Y = ABCD$; (*c*) $X = \overline{A}BC\overline{D}$

8.

A	B	Expression
0	0	0
0	1	1
1	0	1
1	1	0

9.

A	B	C	Odd number of ones
0	0	0	0
0	0	1	1
0	1	0	1
0	1	1	0
1	0	0	1
1	0	1	0
1	1	0	0
1	1	1	1

(*Note:* The function plotted on the truth table above is the function used in a binary adder.

Chapter 7

1. Yes, $X = Y$

2. Yes, $U = V$

3. (*c*) = (*d*)

4. Yes, $X = Y$

5. (*a*) as given

   (*b*) $\overline{X} = (\overline{A} + B)C$ or equivalent

(c) $X + A\bar{B}\bar{C} + A\bar{B}C + \bar{A}\bar{B}\bar{C} + \bar{A}B\bar{C} + AB\bar{C}$

(d) $X = (A + B + \bar{C})(A + \bar{B} + \bar{C})(\bar{A} + \bar{B} + \bar{C})$

(e) $\bar{X} = \bar{A}\bar{B}C + \bar{A}BC + ABC$

(f) $\bar{X} = (\bar{A} + B + C)(\bar{A} + B + \bar{C})(A + B + C)(A + \bar{B} + C)(\bar{A} + \bar{B} + C)$

(g) $X(A, B, C) = \Sigma(000, 010, 100, 101, 110)$

(h) $X(A, B, C) = \Sigma(0, 2, 4, 5, 6)$

6. (a) as given

(b) $\bar{Y} = (A + B)(B + C)$ or equivalent

(c) $Y = \bar{A}\bar{B}\bar{C} + \bar{A}\bar{B}C + A\bar{B}\bar{C}$

(d) $Y = (A + \bar{B} + C)(A + \bar{B} + \bar{C})(\bar{A} + \bar{B} + C)(\bar{A} + \bar{B} + \bar{C})(\bar{A} + B + \bar{C})$

(e) $\bar{Y} = \bar{A}B\bar{C} + \bar{A}BC + AB\bar{C} + ABC + A\bar{B}C$

(f) $\bar{Y} = (A + B + C)(A + B + \bar{C})(\bar{A} + B + C)$

(g) $Y(A, B, C) = \Sigma(000, 001, 100)$

(h) $Y(A, B, C) = \Sigma(0, 1, 4)$

7. Standard product form of X.

$X = (A + B + C)$

Chapter 8

1. (a) Factor $B\bar{D}$

$Z = B\bar{D}(\bar{C} + A + \bar{A}C) = B\bar{D}(A + 1) = B\bar{D}$

(b), (c) $Z = B\bar{D}$

2. (a) $Y = \bar{A}BD + BCD + AD(\bar{C} + \bar{B}C)$

$Y = \bar{A}BD + BCD + AD(\bar{C} + \bar{B})$

$Y = \bar{A}BD + BCD + A\bar{C}D + A\bar{B}D$

(b), (c) $Y = AD + BD$

3. $U = \overline{C}D$; 18 diodes saved

4. $V = C + AB$; 11 diodes saved

5. In order that terms can be combined they have to differ in exactly one variable. One term has n ones (variable = 0); other term has $(n+1)$; differs in one variable.

6. Yes. If no other way, it can be rewritten in standard sum form and terms combined.

Chapter 9

1. $X = \overline{A}\overline{B}\overline{C}\overline{D} + \overline{A}BD + A\overline{C}D + AC\overline{D} + AB\overline{C}$
   or
   $X = \overline{A}\overline{B}\overline{C}\overline{D} + \overline{A}BD + A\overline{C}D + AC\overline{D} + AB\overline{D}$

2. $X = AB\overline{C} + \overline{A}BD + BC\overline{D}$
   or
   $X = B\overline{C}D + \overline{A}BC + AB\overline{D}$

3. Minimized form of X:
   $X = \overline{A}B\overline{C} + A\overline{C}D + \overline{B}C$
   Circuit: 11 diodes, 3 inverters

4. Minimized form of X:
   $X = AB\overline{C} + BD + CD$
   Circuit: 10 diodes, 1 inverter

5.

A	B	C	D	Z
0	0	0	0	1
0	0	0	1	0
0	0	1	0	0
0	0	1	1	0
0	1	0	0	1
0	1	0	1	1
0	1	1	0	0
0	1	1	1	0
1	0	0	0	1
1	0	0	1	1
1	0	1	0	1
1	0	1	1	0
1	1	0	0	1
1	1	0	1	1
1	1	1	0	1
1	1	1	1	1

$Z = AB + A\overline{D} + A\overline{C} + B\overline{C} + \overline{C}\overline{D}$
Circuit: 15 diodes, 2 inverters

Chapter 10

1. (*a*) $U(A, B, C, D) = A\overline{B} + A\overline{C} + BD$

CD \ AB	00	01	11	10
00			1	1
01		1	1	1
11		1	1	1
10				1

(*b*) $V(A, B, C) = (000, 010, 111)$

BC \ A	0	1
00	1	
01		
11		1
10	1	

(*c*) $W(A, B, C, D) = \Sigma(0, 2, 4, 5, 9, 12, 15)$

CD \ AB	00	01	11	10
00	1	1	1	
01		1		1
11			1	
10	1			

(*d*) $X = (A + \overline{B})C + (B + \overline{D})A$

CD \ AB	00	01	11	10
00			1	1
01			1	
11	1		1	1
10	1		1	1

(e) $Y = (A + B)(\overline{A} + \overline{B})$

$B \backslash A$	0	1
0		1
1	1	

(f) $\overline{Z} = (\overline{A} + C + \overline{D})(A + \overline{B} + C)(\overline{C} + D)$

$Z = A\overline{C}D + \overline{A}B\overline{C} + C\overline{D}$

$CD \backslash AB$	00	01	11	10
00		1		
01		1	1	1
11				
10	1	1	1	1

2. (a)

D_1	C_1	D_2	C_2	X
0	0	0	0	0
0	0	0	1	0
0	0	1	0	0
0	0	1	1	1
0	1	0	0	0
0	1	0	1	1
0	1	1	0	0
0	1	1	1	1
1	0	0	0	0
1	0	0	1	0
1	0	1	0	0
1	0	1	1	1
1	1	0	0	1
1	1	0	1	0
1	1	1	0	1
1	1	1	1	0

(b) $X = \overline{D}_1\overline{C}_1 D_2C_2 + \overline{D}_1C_1\overline{D}_2C_2 + \overline{D}_1C_1D_2C_2 + D_1\overline{C}_1D_2C_2 +$
$D_1C_1\overline{D}_2\overline{C}_2 + D_1C_1D_2\overline{C}_2$

(c) $X = \overline{D}_1C_1C_2 + D_1C_1\overline{C}_2 + \overline{C}_1C_2D_2$

(d)

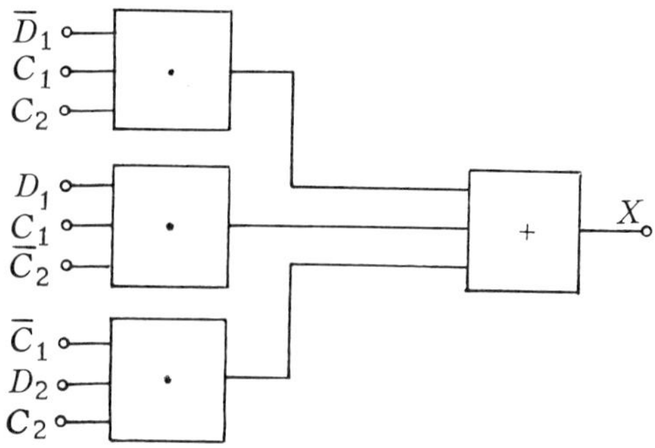

(e)

D_2C_2 \ D_1C_1	00	01	11	10
00			1	
01		1		
11	1	1		1
10			1	

3.

C \ D	0	1
0	1	
1		1

Z

4.

A	B	C	D	X
0	0	0	0	0
0	0	0	1	1
0	0	1	0	0
0	0	1	1	0
0	1	0	0	0
0	1	0	1	0
0	1	1	0	0
0	1	1	1	0
1	0	0	0	0
1	0	0	1	0
1	0	1	0	0
1	0	1	1	1
1	1	0	0	0
1	1	0	1	0
1	1	1	0	0
1	1	1	1	1

CD \ AB	00	01	11	10
00				
01	1			
11			1	1
10				

X

5.

CD \ AB	00	01	11	10
00				
01				1
11		1		1
10				

X

6.

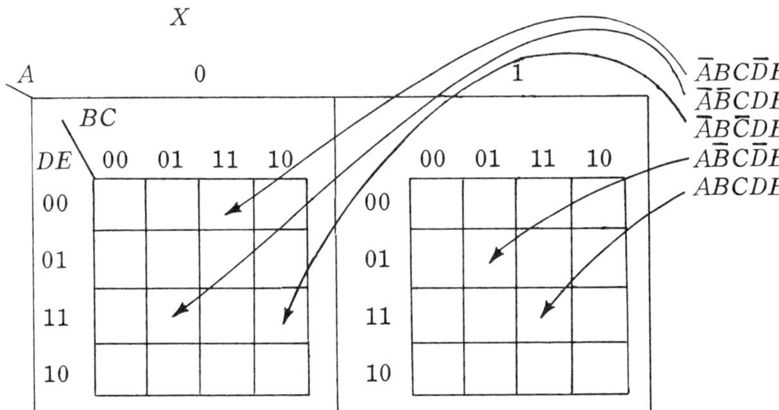

Chapter 11

1. $Z = B\bar{D}$

2. $Y = A\bar{B}\bar{C} + BD$

3. (*a*) $U = \bar{C}D$ (18 diodes saved)
   (*b*) $V = C + AB$

4. (*a*) For clarity, three maps are used:

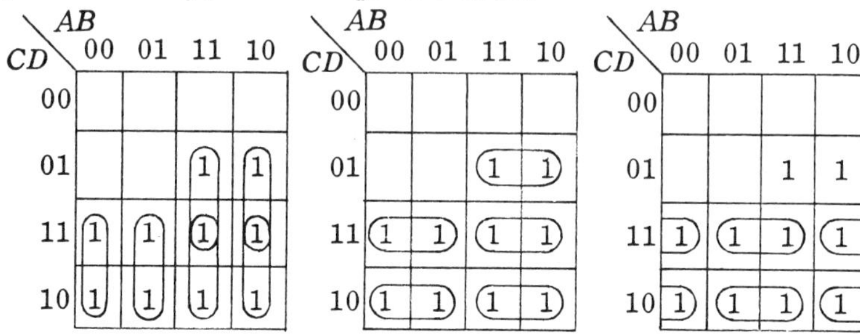

(*b*) For clarity, three maps are used:

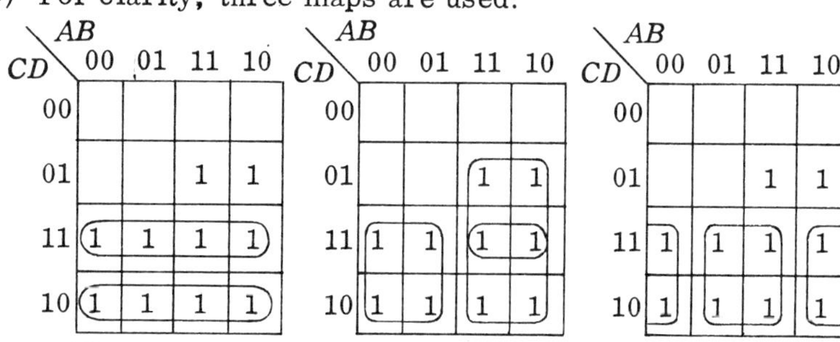

(*c*)

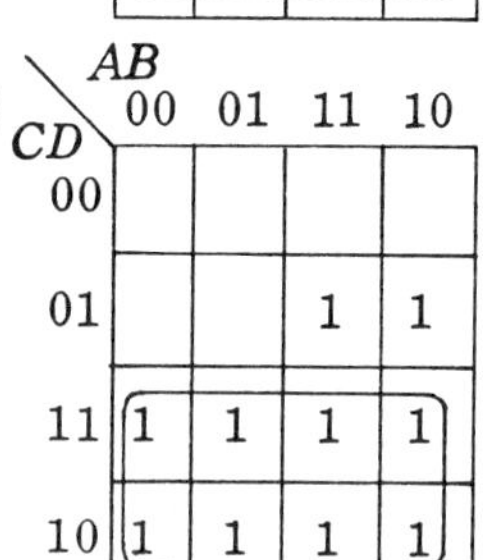

5. (*a*) $Z = \bar{C}D + \bar{A}B\bar{C} + ABC$; (*b*) loop ABD

6. BCD partly covered by loop (prime implicant) $\overline{A}BC$; the rest is covered by AD.

7. $Z = AB + A\overline{D} + A\overline{C} + B\overline{C} + \overline{C}\overline{D}$
Circuit: 15 diodes, 2 inverters

8.

	A	A	$\overline{A}$	$\overline{A}$	
B	12	14	6	4	$\overline{D}$
B	13	15	7	5	D
$\overline{B}$	9	11	3	1	D
$\overline{B}$	8	10	2	0	$\overline{D}$
	$\overline{C}$	C	C	$\overline{C}$	

9. $X = BD + \overline{A}\overline{C}\overline{D} + ACD$

Chapter 12

1. $X = (B + D)(A + \overline{B} + \overline{D})(\overline{A} + \overline{B} + \overline{C})(\overline{A} + B + C)$

2. (*a*) Inverted form: $\overline{X} = \overline{A}\overline{D} + \overline{B}\overline{C} + C\overline{D} + \overline{A}BC$
(*b*) True form: $Y = \overline{B}D + \overline{A}B\overline{D} + A\overline{B}\overline{C}$
(*c*) Inverted form: $\overline{Z} = \overline{A}B\overline{C} + ABD + \overline{A}C\overline{D}$

3. (*a*) Product form; (*b*) sum form; (*c*) product form.

4.

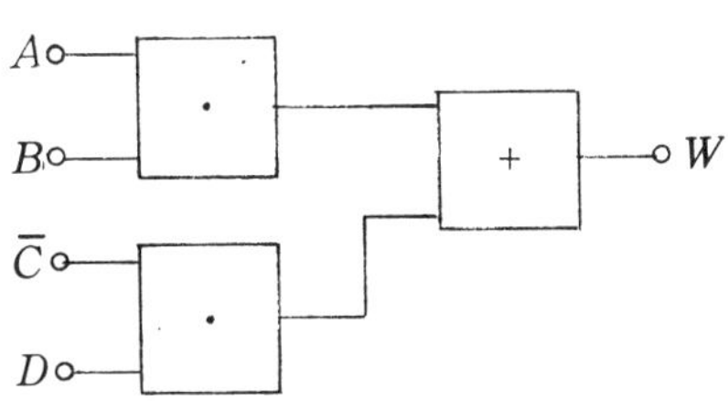

5. (*a*)

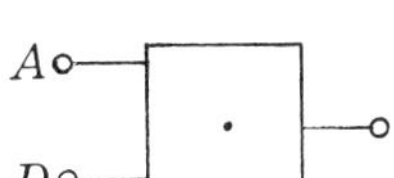

(b)

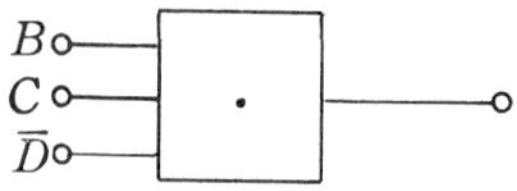

6.

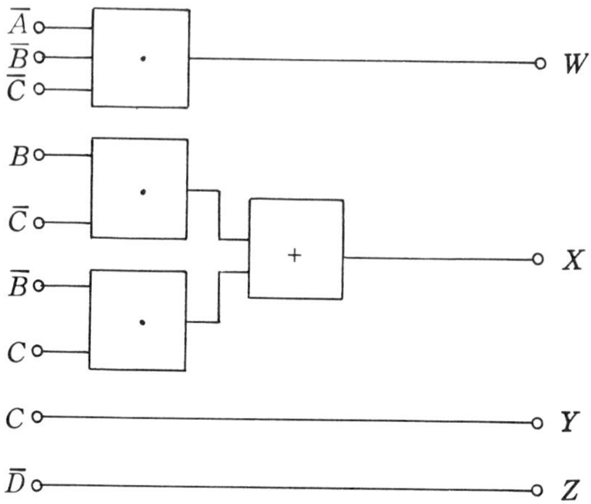

7. (a) $\overline{C}D + AD$; (b) $\overline{A}B\overline{C} + ABC$

Chapter 13

1.

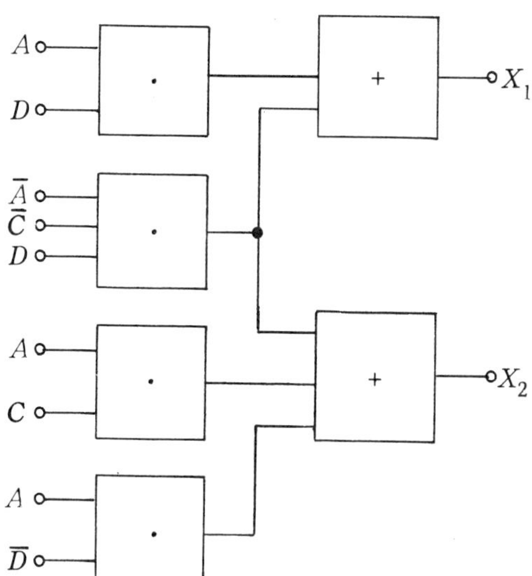

2.

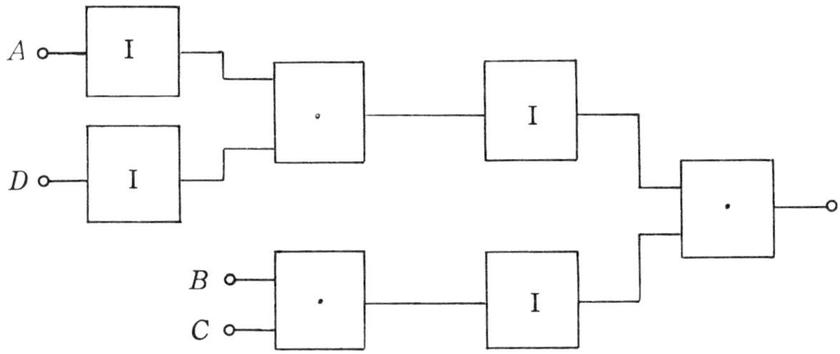

3. $X = \bar{A}B + A\bar{D} = \overline{(A + \bar{B})} + \overline{(\bar{A} + D)}$

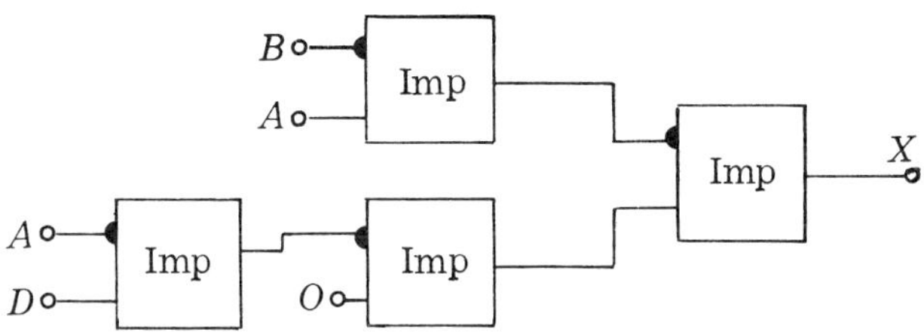

4. To be functionally complete, the combination must perform the three operations AND, OR, and INVERT.

   (*a*) OR block is available.
   (*b*) Any variable or function can be INVERTed by EXCLUSIVE ORing that variable or function with the constant 1.
   (*c*) The AND of two or more variables or functions can be accomplished by INVERTing the OR of the inverse of each variable or function.

REFERENCES

The references in this book are arranged in a slightly different manner. Instead of stating the references in the book itself, the most popular books were chosen and compared to this book. In the table below the chapters of this book are listed with the pages of the reference books which best correspond to the material covered in this book. It should be noted that this list is not complete but represents only those books which seem to be the most popular ones.

In the table these books are listed by the last name of the senior author. Here are the complete references:

Caldwell, Samuel: *Switching Circuits and Logic Design*, John Wiley & Sons, Inc., New York, 1958.

Maley, Gerald A., and John Earl: *The Logic Design of Transistor Digital Computers*, Prentice-Hall, Inc., Englewood Cliffs, N. J., 1963.

Marcus, Michell P.: *Switching Circuits for Engineers*, Prentice-Hall, Inc., Englewood Cliffs, N. J., 1962.

Phister, Montgomery: *Logical Design of Digital Computers*, John Wiley & Sons, Inc., New York, 1958.

Richards, R. K.: *Arithmetic Operations in Digital Computers*, D. Van Nostrand Company, Inc., Princeton, N. J., 1955.

Topics	Chapter numbers in this book	Corresponding pages in other books				
		Richards	Phister	Caldwell	Marcus	Maley
General	1		1–12	1–10		
Basic operations	2		21–25	313–348	46–55	1–12
Theorems	3, 4	26–43		34–118	1–24	35–45
Truth tables	3		30–47	73–77	1–4	35–45 59–69
Inversion	5					
Forms of expressions	7				25–32	52–55
Quine–McCluskey method and prime-implicant chart	7–9	51–66	61–75	145–168	70–84	74–76 92–97
Karnaugh map (or equivalent)	10, 11	66–70	81–96	119–145	85–99	45–55 76–92
Don't cares	12	9	97–105		71	97
Multiple output	13		106–108	174–199 353–363		97–108
Factoring	13			169–171 348–353	99–100	108–113
Binary number system	Appendix B	3	16–20	663–665 673–674	142–146	15–18
Electronic circuits	Appendix C		21–25	319–348	46–55	1–12
Relay logic	Appendix D	39–44		11–33	56–65	

(In the Richards column, 26–43 is bracketed to cover Basic operations through Inversion; in the Phister column, 30–47 is bracketed to cover Theorems through Inversion.)

INDEX

DECIMAL TO BINARY CONVERSION TABLE

Decimal number	Binary number	Index	Decimal number	Binary number	Index
0	0	0	32	100000	1
1	1	1	33	100001	2
2	10	1	34	100010	2
3	11	2	35	100011	3
4	100	1	36	100100	2
5	101	2	37	100101	3
6	110	2	38	100110	3
7	111	3	39	100111	4
8	1000	1	40	101000	2
9	1001	2	41	101001	3
10	1010	2	42	101010	3
11	1011	3	43	101011	4
12	1100	2	44	101100	3
13	1101	3	45	101101	4
14	1110	3	46	101110	4
15	1111	4	47	101111	5
16	10000	1	48	110000	2
17	10001	2	49	110001	3
18	10010	2	50	110010	3
19	10011	3	51	110011	4
20	10100	2	52	110100	3
21	10101	3	53	110101	4
22	10110	3	54	110110	4
23	10111	4	55	110111	5
24	11000	2	56	111000	3
25	11001	3	57	111001	4
26	11010	3	58	111010	4
27	11011	4	59	111011	5
28	11100	3	60	111100	4
29	11101	4	61	111101	5
30	11110	4	62	111110	5
31	11111	5	63	111111	6

THEOREMS OF BOOLEAN ALGEBRA: SUMMARY

OR:	$0 + 0 = 0$	AND:	$0 \cdot 0 = 0$	INVERTER:	$\overline{1} = 0$
	$0 + 1 = 1$		$0 \cdot 1 = 0$		$\overline{0} = 1$
	$1 + 0 = 1$		$1 \cdot 0 = 0$		
	$1 + 1 = 1$		$1 \cdot 1 = 1$		$\overline{\overline{A}} = A$

If $A \neq 1$ then $A = 0$
If $A \neq 0$ then $A = 1$

Sum Form	*Product Form*
$A + 0 = A$	$A \cdot 1 = A$
$A + 1 = 1$	$A \cdot 0 = 0$
$A + A = A$	$A \cdot A = A$
$A + \overline{A} = 1$	$A \cdot \overline{A} = 0$
$A + B = B + A$	$AB = BA$
$A + AB = A$	$A(A + B) = A$
$A\overline{B} + B = A + B$	$(A + \overline{B})B = AB$
$A + B + C = (A + B) + C$	$ABC = (AB)C$
$AB + A\overline{B} = A$	$(A + B)(A + \overline{B}) = A$
$AB + AC = A(B + C)$	$(A + B)(A + C) = A + BC$
$AB + BC + \overline{A}C = AB + \overline{A}C$	$(A + B)(B + C)(\overline{A} + C) = (A + B)(\overline{A} + C)$
$\overline{(A + B + C + \ldots)} = \overline{A}\,\overline{B}\,\overline{C} \ldots$ *	$\overline{ABC \ldots} = \overline{A} + \overline{B} + \overline{C} + \ldots$ *

CD \ AB	00	01	11	10
00	0	4	12	8
01	1	5	13	9
11	3	7	15	11
10	2	6	14	10

Four-variable map

A = 0

DE \ BC	00	01	11	10
00	0	4	12	8
01	1	5	13	9
11	3	7	15	11
10	2	6	14	10

A = 1

DE \ BC	00	01	11	10
00	16	20	28	24
01	17	21	29	25
11	19	23	31	27
10	18	22	30	26

Five-variable map

*These theorems are forms of De Morgan's theorems.

```
float a,b,c;      /* INPUT */
double x1, x2, d; /* OUTPUT */

printf("The roots are: %f %f", x1, x2);
```